Lecture Notes in Electrical Engineering

Volume 1587

The book series *Lecture Notes in Electrical Engineering* (LNEE) publishes the latest developments in Electrical Engineering—quickly, informally and in high quality. While original research reported in proceedings and monographs has traditionally formed the core of LNEE, we also encourage authors to submit books devoted to supporting student education and professional training in the various fields and applications areas of electrical engineering. The series cover classical and emerging topics concerning:

- Communication Engineering, Information Theory and Networks
- Electronics Engineering and Microelectronics
- Signal, Image and Speech Processing
- Wireless and Mobile Communication
- Circuits and Systems
- Energy Systems, Power Electronics and Electrical Machines
- Electro-optical Engineering
- Instrumentation Engineering
- Avionics Engineering
- Control Systems
- Internet-of-Things and Cybersecurity
- Biomedical Devices, MEMS and NEMS

For general information about this book series, comments or suggestions, please contact leontina.dicecco@springer.com.

To submit a proposal or request further information, please contact the Publishing Editor in your country:

China

Jasmine Dou, Editor (jasmine.dou@springer.com)

India, Japan, Rest of Asia

Swati Meherishi, Editorial Director (Swati.Meherishi@springer.com)

Southeast Asia, Australia, New Zealand

Ramesh Nath Premnath, Editor (ramesh.premnath@springernature.com)

USA, Canada

Michael Luby, Senior Editor (michael.luby@springer.com)

All other Countries

Leontina Di Cecco, Senior Editor (leontina.dicecco@springer.com)

**** This series is indexed by EI Compendex and Scopus databases. ****

Gang Mu · Zhengxiang Song · Zhiming Ding ·
Li Han
Editors

Proceedings of the 2nd Electrical Artificial Intelligence Conference, Volume 1

EAIC 2025, 3–5 November, Nanjing, China

Springer

Editors
Gang Mu
Northeast Electric Power University
Changchun Road No.169, Jilin City
Jilin Prov., China

Zhiming Ding
Institute of Software
Chinese Academy of Sciences
Beijing, China

Zhengxiang Song
Xi'an Jiaotong University
Xi'an, Shaanxi, China

Li Han
Institute of Electrical Engineering
Chinese Academy of Sciences
Beijing, China

ISSN 1876-1100 ISSN 1876-1119 (electronic)
Lecture Notes in Electrical Engineering
ISBN 978-981-95-7931-0 ISBN 978-981-95-7932-7 (eBook)
https://doi.org/10.1007/978-981-95-7932-7

This Springer imprint is published by the registered company Springer Nature Singapore Pte Ltd.
The registered company address is: 152 Beach Road, #21-01/04 Gateway East, Singapore 189721, Singapore

Contents

Integrated Energy Base Scheduling in Demand Response Scenarios: A Collaborative Model of Graph Attention Network and SAC Algorithm

Jiaqi Zhao, Yanxia Ma, Na Chen, Zhiyuan Wang, Gang Li, and Ruimin Liu

Abstract In the scheduling of integrated energy systems, the dynamic supply and demand matching and multi-energy coupling problems in demand response scenarios need to be solved urgently. This study proposes a collaborative model of graph attention network (GAT) and SAC algorithm. First, GAT is used to extract energy node characteristics and construct energy network topological relationships, and the multi-head attention mechanism is used to quantify the coupling weights of multi-energy flows such as electricity, heat, and gas; then the SAC algorithm is used to construct a dynamic decision-making framework, and the entropy regularization strategy is used to achieve multi-objective collaborative optimization of supply and demand balance, cost optimization, and renewable energy consumption. A composite reward function is designed that includes energy cost, carbon emission penalty, and load fluctuation rate. The study confirms that the GAT-SAC model can effectively capture the spatiotemporal characteristics of the energy network, providing a new method for solving dynamic energy scheduling problems.

Keywords IES scheduling · Demand response · Graph attention network (GAT) · Soft actor–critic (SAC) · Multi-energy coupling

1 Introduction

With the access of a high proportion of renewable energy and the growth of diversified load demand, the scheduling of integrated energy systems faces core challenges such as insufficient accuracy in dynamic supply and demand matching and complex multi-energy flow coupling mechanisms. Traditional physical model-driven methods

J. Zhao · Y. Ma · N. Chen
State Grid Ningxia Electric Power Economic Technology Research Institute, Ningxia, China

Z. Wang (✉) · G. Li · R. Liu
Nanjing Normal University, Nanjing, China
e-mail: xhl110302@126.com

G. Mu et al. (eds.), *Proceedings of the 2nd Electrical Artificial Intelligence Conference, Volume 1*, Lecture Notes in Electrical Engineering 1587,
https://doi.org/10.1007/978-981-95-7932-7_1

are limited by rigid topological assumptions and linear approximations, making it difficult to characterize the nonlinear interactive characteristics of heterogeneous energy networks; while data-driven methods have environmental adaptability, they are separated from feature extraction and strategy optimization, resulting in the loss of multi-time and space scale correlation information.

Existing research has made progress in multi-energy complementary scheduling modeling, prediction technology and new algorithm application. Huang et al. [1] proposed a multi-energy complementary optimization scheduling model integrating multiple objectives and multiple time scales, and using the branch cutting method to solve under the condition of meeting safety constraints. Wang et al. [2], Huang et al. [3], Liu et al. [4], Qiao and Gao [5] established an electricity price prediction model based on long short-term memory neural network (LSTM) and a multi-energy complement day-ahead optimization scheduling model, achieve the day-ahead optimization scheduling goal of maximizing the comprehensive benefits of the energy base. Chen et al. [6] proposed a blockchain consensus mechanism specifically designed to support mathematical optimization problems, called proof of solution (PoSo). Lei et al. [7] and Guo et al. [8] focused on a deep reinforcement learning (DRL)-based energy dispatch approach for IoT-driven smart isolated microgrids equipped with diesel generators (DG), photovoltaic (PV) panels, and batteries. Current methods mostly rely on physical model simplification or single-objective optimization, fail to fully consider the temporal and spatial correlation and uncertainty interaction of multi-energy networks, lack an end-to-end scheduling framework for joint feature learning and dynamic decision-making, and restrict the adaptive optimization capabilities in complex scenarios.

This paper proposes a collaborative model of graph attention network (GAT) and flexible actor-critic (SAC), aiming to break through the inherent paradigm of hierarchical processing of "feature representation-decision optimization". This method provides a unified solution for spatiotemporal correlation modeling and real-time decision-making in heterogeneous energy systems.

2 Method

2.1 GAT Feature Extraction Module

The input of the GAT module is a node feature matrix composed of multi-source heterogeneous data in the integrated energy system, covering key parameters such as electricity, heat, and gas load demand, real-time output of renewable energy (wind power, photovoltaic), and charging and discharging status of energy storage equipment. The GAT graph attention layer structure is as follows (Fig. 1).

To characterize the multi-energy coupling relationship, the system is modeled as a heterogeneous graph, which contains three types of nodes (generators, energy storage, and loads) and two types of directed edges (energy transmission paths and

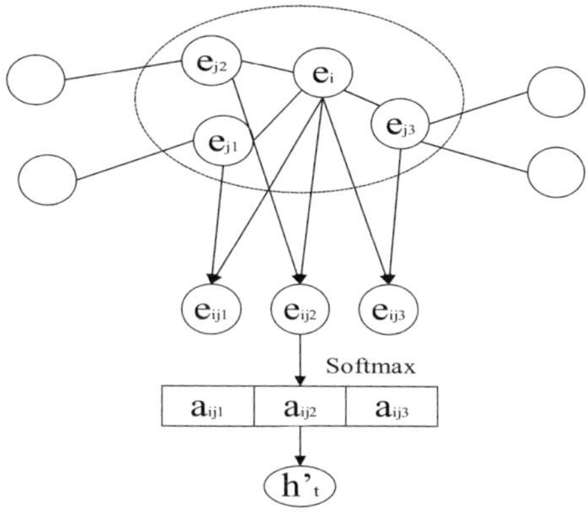

Fig. 1 Structure of graph attention layer

multi-energy conversion links). Specifically, the power network edge represents the connection between the substation and the transmission line, and its attributes include transmission capacity (MW) and impedance (Ω); the thermal network edge includes the heating radius (km) and heat loss coefficient (W/m K); and the gas network edge defines the pipeline gas transmission capacity (m^3/h) and pressure drop parameters. The topological structure is jointly encoded by the adjacency matrix A and the edge type matrix E, where A_{ij} indicates that there is energy interaction between nodes i and j, and E_{ij} identifies the physical type of the edge. This modeling method explicitly distinguishes the direction of energy flow and the level of conversion, providing a structured prior for the attention mechanism.

Based on the topological graph, GAT uses a multi-head attention mechanism (8 heads) to dynamically quantify the coupling strength between nodes [9]. For node i and its neighbor $j \in N(i)$, the feature vectors h_i and h_j are first mapped to the query $q_i = W_q h_i$, key $k_j = W_k h_j$ and value $v_j = W_v h_j$ through the shared weight matrix W_k, and then the attention coefficient is calculated:

$$a_{ij} = soft\max\left(\frac{q_i^T k_j}{\sqrt{d_k}} + \Phi\left(E_{ij}\right)\right) \tag{1}$$

d_k is the key vector dimension and $\Phi\left(E_{ij}\right)$ is the edge type embedding function (learnable vector lookup table), which is used to encode the physical constraints of different energy flows. Multi-head attention executes the above process in parallel, concatenating the outputs of each head and fusing them through the fully connected layer to obtain the updated features of node i:

$$h_i' = \sigma\left(\Big\|_{m=1}^{M} \sum_{j \in N(i)} a_{ij}^m v_{ij}^m\right) \tag{2}$$

(M = 8 is the number of heads, sigma is the ELU activation function). This mechanism enables the model to adaptively focus on key coupling relationships (such as prioritizing the adjustment of energy storage nodes when wind power fluctuates) and quantify the intensity of cross-energy interactions (such as dynamic adjustment of the electric-heat conversion coefficient with load changes).

2.2 SAC Optimization Framework

The state space of the SAC algorithm is composed of the dynamic feature vector (128 dimensions) output by the GAT module and external real-time variables (5 dimensions such as electricity price, temperature, wind speed, etc.), forming a 133-dimensional continuous state vector. Among them, the GAT feature vector encodes the topological correlation of the energy network, the real-time electricity price reflects the fluctuation of the electricity market, and the meteorological data (wind speed, light intensity) drives the forecast correction of renewable energy output. To ensure the feasibility of the action, the Tanh activation function is used to constrain the output range, and the out-of-limit action is verified in real time through environmental feedback [10]. The reward function is designed as a multi-objective weighted sum:

$$R_t = 0.35R_{cost} + 0.25R_{carbon} + 0.4R_{load} - \lambda R_{penalty} \tag{3}$$

Among them, R_{cost} is the operating cost item, including fuel costs, operation and maintenance costs (equipment start-up and shutdown losses and demand response compensation costs); R_{carbon} is the carbon emission penalty item, which calculates the CO_2 emissions by unit type (gas-fired unit, coal-fired unit) and multiplies it by the carbon trading price; R_{load} is the load volatility reward, which calculates the negative value of the root mean square error (RMSE) between the actual load and the predicted value; $R_{penalty}$ is a safety constraint penalty term.

SAC enhances the randomness of the strategy by maximizing entropy, and its optimization objective is expanded to:

$$J(\pi) = IE_{(s,a)\sim\rho\pi}\left[Q(s,a) + \alpha H(\pi(\cdot|s))\right] \tag{4}$$

$Q(s,a)$ is the state-action value function; $H(\pi)$ is the entropy of the policy distribution; α is the adaptive temperature coefficient (initial value 0.2, dynamically adjusted during training). The policy network uses the reparameterization technique to generate Gaussian distribution actions and updates parameters through gradient ascent to improve the expected return [11]. The double Q network design suppresses overestimation by taking the minimum value:

$$Q_{t\arg et} = \min_{i=1,2} Q_{\Phi i}(s,a) - \alpha \log \pi(a|s) \tag{5}$$

During training, the experience replay buffer randomly samples batches of data (batch size 256) and updates the Q network by minimizing the temporal difference error (TD error):

$$L(\Phi_i) = IE_{(s,a,r,s')}\left[\left(Q_{\Phi i}(s,a) - \left(r + \gamma IE_{a' \sim \pi} Q_{t\,\mathrm{arg}\,rt}\left(s', a'\right)\right)\right)^2\right] \tag{6}$$

The entropy coefficient is automatically adjusted through gradient descent to maintain the target entropy. This mechanism enables the strategy to dynamically balance between exploration (trying new actions of multi-energy supply) and utilization (prioritizing high-return actions), effectively approaching the multi-objective Pareto optimal solution set.

2.3 GAT-SAC Synergy Mechanism

The GAT module focuses on extracting the spatial correlation characteristics of energy networks, and dynamically quantifies the complementary potential of multi-energy flows between regions (such as the power support capacity of wind, solar and storage clusters, and the conversion efficiency of electric, thermal and gas coupling nodes) through graph attention weights. The SAC algorithm generates time series dispatch instructions based on the spatiotemporal feature vectors (including topological correlation strength, supply and demand margin, etc.) output by GAT, combined with external market signals. The two work together to break through the limitations of the traditional method of "feature extraction-decision optimization" staged processing, allowing the dispatch strategy to adapt to both spatial heterogeneity and temporal volatility.

The training process adopts an integrated design, and the parameters of the GAT and SAC networks are optimized synchronously through back propagation. GAT processes energy node data to generate feature vectors, which are input into the SAC policy network to generate action instructions; the action instructions act on the simulation environment to obtain the next state and reward signal [12]. Table 1 shows the coordinated changes of key parameters during the joint training process.

Table 1 Co-variation of key parameters during joint training

Training epoch	Gat attention head weight variance	SAC policy entropy	Feature vector similarity	Q-value loss	Gradient coupling strength
0	0.12	1.58	0.31	8.72	0.05
200	0.28	1.02	0.67	4.15	0.18
400	0.35	0.76	0.82	2.33	0.24
600	0.41	0.63	0.89	1.47	0.31
800	0.39	0.59	0.91	1.05	0.28

The decrease in the entropy value of the SAC strategy indicates that the decision-making has shifted from exploration to stable optimization, and the improvement in the similarity of feature vectors verifies the effectiveness of spatial correlation learning.

3 Results and Discussion

3.1 Simulation Settings

The simulation environment is built based on Python 3.8 + PyTorch 1.9, and the Gurobi 9.5 solver is integrated to process mixed integer programming constraints. The test case uses the IEEE 39-node power system to expand the multi-energy network: the power subsystem includes 39 generators (including 12 wind turbines and 8 photovoltaic power stations), 24 energy storage devices (50% each of lithium batteries and pumped storage); the thermal subsystem is connected to 16 regional heating stations and 9 ground source heat pump links; the gas subsystem includes 7 gas source stations and 12 gas turbines. The simulation time span is 8760 h throughout the year, with a time resolution of 15 min. The real-time electricity price adopts the CAISO market time-of-use electricity price mechanism, and the peak-to-valley price difference ratio is set to 3:1.

The experiment selects the deep reinforcement learning baseline DQN (fully connected network hidden layer 256 × 128) and the proximal policy optimization PPO (Actor-Critic structure, KL divergence threshold 0.01) as the comparison method, and uses the same hyperparameters as GAT-SAC: learning rate 0.001 (Adam optimizer), discount factor 0.99, experience replay cache capacity 1e6, batch size 256. All methods are run on the same training set (first 300 days) and test set (last 65 days). Each training cycle contains 1000 rounds, and the single duration is 24 h (96 decision steps).

3.2 Testing Process

First, before the start of daily simulation, the environment is reset to the initial state (energy storage SOC = 50%, load baseline value), and a scheduling decision is made every 15 min (a total of 96 steps/day). The timestamp of the action instruction generation of each method (the calculation delay from state input to action output), the real-time operation cost (fuel cost + operation and maintenance fee + demand response compensation) and the actual renewable energy consumption (wind power/photovoltaic output minus the abandoned power) are recorded. After completing the full cycle test, the average daily operation cost, the average renewable energy consumption rate and the average instruction delay of each method are calculated.

Finally, the performance differences of GAT-SAC, DQN and PPO in three dimensions are compared. All tests are repeated multiple times to eliminate randomness, and the results are taken with a 95% confidence interval.

3.3 *Results Analysis*

Figures 2 and 3 show the test results of daily average operating cost and renewable energy acceptance rate. The average daily operating cost of GAT-SAC ranges from 3,609 yuan (Test 7) to 3,998 yuan (Test 10), with 11 test values below 3,800 yuan and the lowest value 389 yuan lower than its own highest value; the cost fluctuation range of DQN is larger, with the lowest value being 3,869 yuan (Test 11) and the highest value being 4,249 yuan (Test 17), and half of the test values exceeding 4,100 yuan; the cost span of PPO is the largest, ranging from 3,918 yuan (Test 16) to 4,266 yuan (Test 9), with 10 test values above 4,100 yuan. In terms of renewable energy absorption rate, GAT-SAC performed stably, with all 20 tests above 89%, with a peak of 91.5% (Test 8 and Test 18) and a minimum of 89.3% (Test 16 and Test 19). The DQN absorption rate ranges from 87.5% (Test 7 and Test 11) to 89.3% (Test 2), with 6 test values below 88%. The PPO absorption rate fluctuates between 87.9% (Test 6 and Test 18) and 89.8% (Test 15), with 11 test values below 89%. Among the three methods, GAT-SAC has the smallest cost range and the highest frequency of high absorption rate ($\geq$90%), DQN presents the characteristics of "medium cost-low absorption", and PPO has the contradiction of "high cost-medium absorption". The data show that the method in this paper performs best between cost control and absorption capacity, and the internal volatility of the results is low.

Table 2 shows the demand response time test results. GAT-SAC has a stable latency of 4.9–5.6 s in all scenarios, mainly due to the real-time compression of energy topology features by its graph attention network, which reduces the amount of input data for the policy network. DQN has a latency of 9.7–11.3 s due to the redundancy of fully connected network parameters, and needs to re-explore the state space when the load changes suddenly, further increasing the computational load. Due to the Actor-Critic dual network structure and KL divergence constraint calculation, the delay of PPO increases to 13.4–16.1 s, especially in multi-energy coupling scenarios, where the feasibility of constraints needs to be frequently verified. Traditional MPC solves mixed integer programming problems online, and the solver iteration time increases exponentially with the complexity of the scenario, which cannot meet the minute-level response requirements. Experiments show that GAT-SAC significantly reduces the computational complexity while ensuring scheduling accuracy through feature embedding dimensionality reduction and policy network structure optimization.

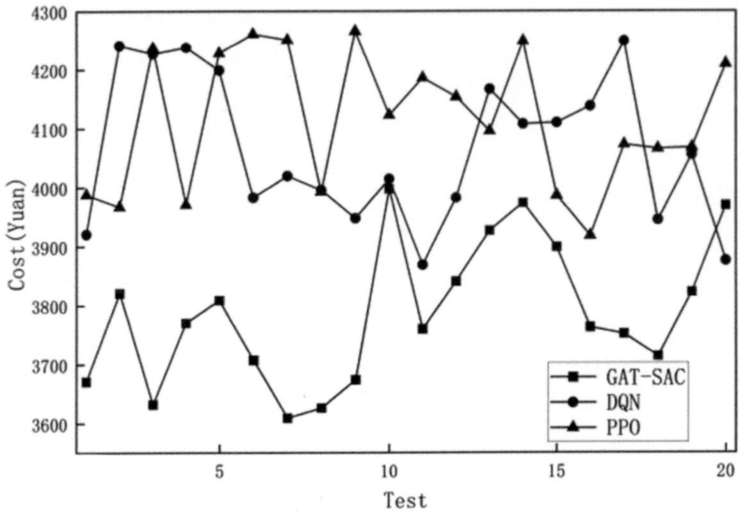

Fig. 2 Average daily operating costs

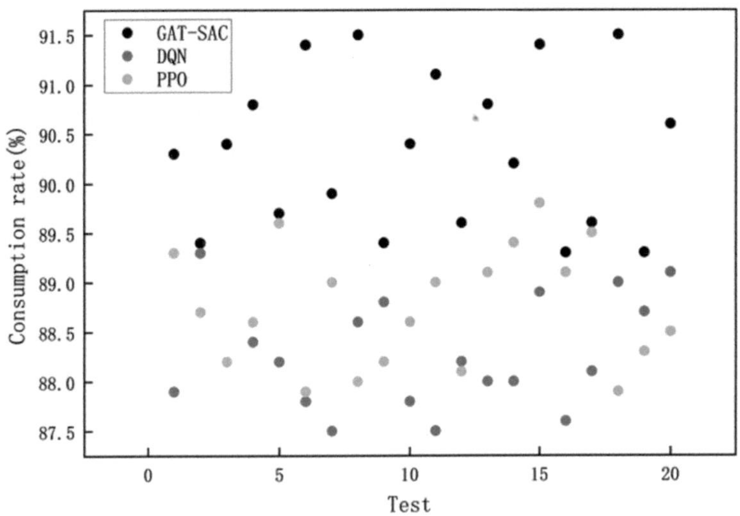

Fig. 3 Renewable resource consumption rate

4 Conclusion

This paper proposes a collaborative scheduling model of graph attention network and flexible Actor-Critic algorithm for demand response scenarios. By integrating energy network topology feature extraction and multi-objective dynamic decision-making mechanism, it breaks through the limitations of traditional methods in

Table 2 Demand response time

Test scenario	GAT-SAC	DQN	PPO	Traditional MPC
Base load fluctuation	4.8	9.1	12.3	32.5
Sudden drop in RES output	5.2	10.6	14.8	41.2
Multi-energy emergency dispatch	5.6	11.3	16.1	48.7
Real-time price surge	4.9	9.7	13.4	35.1
ESS SOC critical alarm	5.1	10.2	15.0	38.9

complex multi-energy flow coupling and real-time response efficiency. The GAT module effectively captures the spatiotemporal correlation characteristics between electricity, heat, and gas multi-energy nodes, and solves the problem of dynamic interaction modeling of heterogeneous energy; the SAC algorithm combines the entropy regularization strategy to achieve multi-objective collaborative optimization of economy, environmental protection, and system stability. In the future, it can be extended to scenarios such as multi-time scale rolling optimization and cross-regional collaborative dispatch.

Acknowledgements This work was supported by the Ningxia Natural Science Foundation (No. 2024AAC03762).

References

1. Huang Y, Hao F, Teng J, Jie K, Chen G, Xia K (2024) Research on multi-objective and multi-time scale optimization dispatch of multi-energy complementary energy base. Inner Mongolia Electr Power Technol 42(4):32–41
2. Wang Y, Liu S, Li T, Zhou Y, Wang X (2023) Comparative study on short-term dispatch operation mode of water-wind-solar complementary energy base in Yalong River energy base. J Hydraul Eng 54(4):439–450
3. Huang X, Gesang Y, Wu Z, Yan S (2022) Multi-time scale optimization scheduling. Hydropower 48(1):106–111
4. Liu N, Kang J, Zhao C (2024) Development model of multi-energy complementary energy base and comprehensive benefit improvement method. Proc CSEE 44(4):1339–1351
5. Qiao Y, Gao R (2024) Research on model construction and capacity optimization configuration of multi-energy complementary clean energy base. People's Yellow River 46(S01):97–98+101
6. Chen S, Mi H, Ping J et al (2022) A blockchain consensus mechanism that uses Proof of Solution to optimize energy dispatch and trading. Nat Energy 7(6):495–502
7. Lei L, Tan Y, Dahlenburg G et al (2020) Dynamic energy dispatch based on deep reinforcement learning in IoT-driven smart isolated microgrids. IEEE Internet Things J 8(10):7938–7953
8. Guo Z, Wei W, Shahidehpour M et al (2022) Optimization methods for dispatch and control of energy storage with renewable integration. IET Smart Grid 5(3):137–160
9. Mock JW, Muknahallipatna SS (2023) A comparison of ppo, td3 and sac reinforcement algorithms for quadruped walking gait generation. J Intell Learn Syst Appl 15(1):36–56
10. Wu Z, Pan S, Chen F et al (2020) A comprehensive survey on graph neural networks. IEEE Trans Neural Netw Learn Syst 32(1):4–24

11. Dwivedi VP, Joshi CK, Luu AT et al (2023) Benchmarking graph neural networks. J Mach Learn Res 24(43):1–48
12. Yuan H, Yu H, Gui S et al (2022) Explainability in graph neural networks: a taxonomic survey. IEEE Trans Pattern Anal Mach Intell 45(5):5782–5799

Graph Convolutional Network-Based Deep Reinforcement Learning for Dynamic Distribution System Reconfiguration

Ruilin Wang

Abstract Distribution network reconfiguration (DNR) is a key measure for enhancing the operational economy and reliability of power grids. However, traditional optimization methods rely on accurate models and are computationally time-consuming, making them difficult to apply in real-time scenarios. To address this, this paper proposes a DNR method based on Graph Convolutional Networks (GCN) and Deep Q-Learning. The GCN is utilized to extract grid topological information and construct node representations that integrate both global and local features. A reward function incorporating a nonlinear reward factor is designed to enhance the agent's sensitivity to different power losses. Furthermore, an action masking mechanism based on graph theory is introduced to ensure that actions satisfy radial constraints, thereby improving exploration efficiency. Experiments on the IEEE 33-node system demonstrate that the proposed method remains effective in noisy environments, achieving an average power loss reduction of approximately 36.2%, and exhibits good robustness and real-time performance.

Keywords Distribution network reconfiguration · Graph convolutional network · Deep reinforcement learning · Deep Q-network

1 Introduction

The power system, as indispensable infrastructure in modern society, demands high levels of security, stability, and economic operation. As the final segment of the power system that directly interfaces with end-users, the operational efficiency and power supply quality of the distribution network have a direct impact on social production and daily life. In recent years, with the acceleration of the global energy transition and the continuous growth of electricity loads, the structure and operational

R. Wang (✉)
University of Electronic Science and Technology of China, Chengdu, Sichuan, P.R. China
e-mail: wrainy2025@163.com

G. Mu et al. (eds.), *Proceedings of the 2nd Electrical Artificial Intelligence Conference, Volume 1*, Lecture Notes in Electrical Engineering 1587,
https://doi.org/10.1007/978-981-95-7932-7_2

patterns of distribution networks are undergoing significant transformations. Urbanization has led to increasing load density, imposing higher requirements on the power supply capacity and power quality of distribution networks. Against this backdrop, optimizing distribution network performance through effective operational control measures to reduce network losses and enhance operational economy and reliability has become a key focus in both academia and industry.

Distribution network reconfiguration (DNR) stands as one of the critical approaches to addressing the aforementioned challenges. Its primary objective is to optimize system operational indicators by altering the open/closed states of sectionalizing and tie switches, thereby adjusting the network topology. However, DNR is inherently a complex nonlinear, multi-constrained, combinatorial optimization problem. To tackle this complexity, researchers have proposed various methods, such as mixed-integer linear programming [1], genetic algorithms [2, 3], particle swarm optimization [4, 5], and ant colony optimization [6]. Most of these algorithms fall under the category of model-based optimization methods and exhibit certain limitations. They heavily rely on accurate grid model parameters, which are often difficult to guarantee in practical systems. Additionally, these algorithms require extensive objective function evaluations, i.e., power flow calculations, making the computational process time-consuming and challenging to apply in real-time reconfiguration scenarios that demand rapid responses.

In recent years, breakthroughs in deep reinforcement learning (DRL) have introduced a new paradigm for solving complex sequential decision-making problems [7]. DRL agents learn optimal strategies directly from high-dimensional raw data through repeated interactions with the environment, without relying on explicit system models, offering inherent model-free advantages. The Deep Q-Network (DQN), known for its strong capability in handling high-dimensional state spaces, is one such method [8]. However, directly applying standard DRL methods like DQN to distribution network reconfiguration poses significant challenges. Standard DQN typically employs fully connected neural networks to process states, which struggle to effectively capture the complex spatial relationships between nodes and branches. To address this issue, Graph Neural Networks (GNNs) provide an ideal solution. Recent studies have demonstrated the effectiveness of combining graphical representations with DRL for power system control tasks, such as voltage regulation [9]. GNNs are specifically designed to handle non-Euclidean graph-structured data, learning graph embeddings closely related to the graph topology through message-passing mechanisms that aggregate neighbor information.

This paper integrates Graph Convolutional Networks (GCN) with Deep Q-Learning to construct an intelligent agent capable of understanding grid topology information for solving the DNR problem. The main contributions of this work are as follows: A GCN-DQN architecture is developed, leveraging multi-layer GCNs to extract spatial structural features of the distribution network, thereby enriching the informational basis for the Q-function. A reward function incorporating a nonlinear factor is proposed, introducing a relative improvement reward coefficient based on the current power loss level to enhance the agent's sensitivity across varying loss conditions. An action masking mechanism based on graph theory is designed to efficiently filter switch action pairs that comply with the radial constraints of the grid, thereby improving exploration efficiency.

The remainder of this paper is organized as follows: Sect. 2 elaborates on the problem formulation of DNR, including the objective function and constraints. Section 3 provides a detailed introduction to the proposed GCN-DQN framework. Section 4 presents and analyzes the experimental results. Finally, Sect. 5 concludes the paper and outlines future research directions.

2 System Model and Problem Formulation

This section establishes the mathematical and reinforcement learning framework for the distribution network reconfiguration (DNR) problem.

2.1 Mathematical Modeling

The DNR problem is formulated as a constrained optimization problem with the following components.

Objective Function, the primary goal is to minimize active power loss, defined as:

$$\min P_{loss} = \sum_{k=1}^{N_{branch}} R_k |I_k|^2 \tag{1}$$

where R_k and I_k represent the resistance and current of branch k, respectively, and N_{branch} is the total number of branches.

Constraints, any feasible network configuration must satisfy the following operational and topological constraints.

Radiality Constraint: The network topology must remain connected and radial (tree-structured), containing no loops and no isolated nodes. Let N be the number of nodes, b the number of branches, A the node-branch incidence matrix, and x the binary decision vector where $x_i = 1$ indicates branch i is closed. The radiality constraint is expressed as:

$$rank(A \cdot diag(x)) = N - 1 \tag{2}$$

$$\sum_{i=1}^{b} x_i = N - 1 \tag{3}$$

Node Voltage Constraint: The voltage magnitude V_i at each node i must remain within permissible limits:

$$V_{\min} \leq V_i \leq V_{\max} \tag{4}$$

Power Flow Equations Constraints: The power flow equations, which govern the physical behavior of the distribution network, must be satisfied for any feasible configuration. Using the DistFlow model for a radial network, these equations for each branch ij connecting nodes i and j are given by:

$$P_{ij} - p_j - \sum_{k:j \to k} P_{jk} = I_{ij}^2 R_{ij} \tag{5}$$

$$Q_{ij} - q_j - \sum_{k:j \to k} Q_{jk} = I_{ij}^2 X_{ij} \tag{6}$$

$$V_j^2 = V_i^2 - 2(R_{ij} P_{ij} + X_{ij} Q_{ij}) + \left(R_{ij}^2 + X_{ij}^2\right) I_{ij}^2 \tag{7}$$

$$I_{ij}^2 = \frac{P_{ij}^2 + Q_{ij}^2}{V_i^2} \tag{8}$$

2.2 *Reinforcement Learning Formulation*

To solve the DNR problem using deep reinforcement learning, we construct a Markov Decision Process (MDP) that maps the one-shot combinatorial optimization into a sequential decision-making process. The core idea is to decompose the complex search for an optimal topology into a trajectory of sequential switch operations. Each step in this trajectory represents a single, atomic reconfiguration action that incrementally transforms the network topology. The MDP is formally defined by the tuple (S, A, R, γ).

State Space (S): The state s_t observed by the agent at time step t is a comprehensive representation of the grid, including: Vectors of active and reactive power for all branches $(P, Q) \in R^{2 \times N_{branch}}$. A vector indicating the open/closed status of all switches $status \in \{0, 1\}^{N_{branch}}$. A vector of node voltage magnitudes $V \in R^{N_{bus}}$. The network topology represented by an adjacency matrix $A \in \{0, 1\}^{N_{branch} \times N_{branch}}$.

Action Space (A): An action a_t is defined as a paired operation: "close one normally open tie-switch and simultaneously open one normally closed sectionalizing switch." This pairing inherently guides the agent towards exploring feasible topological changes. The theoretical size of this combinatorial action space is $N_{branch} \times N_{branch}$. A key innovation is the use of an action masking mechanism, which filters out actions that would violate the radiality or connectivity constraints, drastically reducing the effective action space and improving learning efficiency.

Reward Function (R): The reward function is meticulously designed to guide the agent toward the optimization objective while ensuring constraint satisfaction. It is formulated as:

$$r = -P_{loss}^{new} + \beta \cdot \frac{P_{loss}^{old} - P_{loss}^{new}}{P_{loss}^{old}} \cdot 10^4 + r_{penalty} \tag{9}$$

The reward r_t consists of three components:

Base Reward $-P_{loss}^{new}$: Provides a continuous signal encouraging the reduction of total power loss relative to the initial state.

Improvement Reward $\beta \frac{P_{loss}^{old} - P_{loss}^{new}}{P_{loss}^{old}}$: Offers an additional positive incentive specifically for actions that successfully reduce losses. The nonlinear factor $\beta = k \ln\left(\frac{P_{loss}^{reference}}{P_{loss}^{old}}\right)$ adjusts the sensitivity of this reward based on the current loss level, ensuring the agent remains motivated even when losses are low.

Penalty Term $r_{penalty}$: A penalty term is applied to exclude invalid or unstable network states, such as non-convergent power flow solutions and bus voltage limit violations.

Episode and Termination: An episode represents one complete reconfiguration task, starting from an initial topology and proceeding through a sequence of switch actions. The episode terminates when a terminal state is reached, defined by one of the following conditions:

The network reaches a stable, low-loss configuration where subsequent actions yield negligible improvement $|\Delta P_{loss}| \leq \varepsilon$. A maximum number of steps $T_{\max}$ is executed, preventing excessively long trajectories. No improving, feasible actions are available, indicating a local optimum.

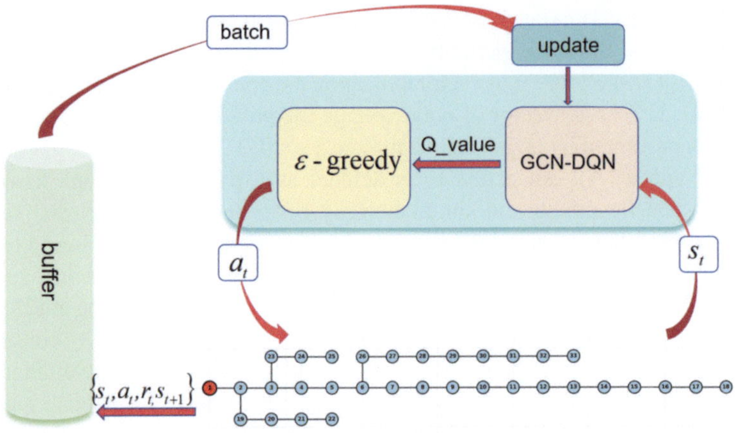

Fig. 1 GCN-DQN agent environment interaction framework diagram

3 GCN-DQN Based Reconfiguration Method

3.1 Overall Framework

The decision-making process of the proposed GCN-DQN method is illustrated in Fig. 1. The agent obtains state information from the environment and outputs Q-values for all possible actions. Actions are selected and executed using an ε-greedy policy. The environment then transitions to a new state and returns a reward. These experiences are stored in a replay buffer for network updates.

3.2 Graph Convolutional Network Design

The GCN generates new, more powerful feature representations for nodes by aggregating features from the node itself and its neighboring nodes.

Graph Construction: The distribution network is modeled as a graph $G = (V, E)$, where nodes V represent buses, and edges E represent branches.

GCN Layer: Considering the physical significance of distribution network branch characteristics, this paper employs an improved graph convolution operation to aggregate topological information. The operation of a single GCN layer can be expressed as:

$$H^{(l+1)} = \sigma\left(AH^{(l)}W_{neighbor}^{(l)} + H^{(l)}W_{self}^{(l)}\right) \tag{10}$$

Here, A is the adjacency matrix of the network, $H^{(l)}$ is the node feature matrix at the l-th layer, $W_{neighbor}$ and W_{self} are the trainable weight matrices for neighbor aggregation and self-loop connection, respectively, and σ is the LeakyReLU activation function.

3.3 Deep Q-Network

The network structure is shown in Fig. 2. First, the branch feature matrix (containing active and reactive power) and the adjacency matrix are input and passed through 4 layers of graph convolution for topological feature extraction. Each GCN layer aggregates neighbor information through the message-passing mechanism defined by Eq. (10). After multiple stacked layers, the feature representation μ of each branch incorporates topological and electrical information from its multi-hop neighbors, enabling global perception of the network state. The system performs global average pooling on the branch features μ to obtain a state vector $h_{global} \in R^d$. This vector is broadcast to each branch and concatenated with μ. Simultaneously, node voltage information $V \in R^{N_{bus}}$ is mapped to the branch dimension via a linear layer and concatenated with the switch status vector into the comprehensive feature tensor, forming a feature tensor that encompasses local, global, topological, and electrical information. This feature tensor is fed into dual action heads. Both the closing action head and the opening action head are multi-layer perceptrons (MLPs) composed of fully connected layers and activation functions. To synthesize the complete action-value function, the Q_{close} and Q_{open} are combined via an outer sum operation, constructing the original Q-value matrix of dimension $N_{branch} \times N_{branch}$, where each element Q_{ij} represents the value of closing branch i and opening branch j. Finally, the masking mechanism is applied: Q-values for illegal operations are set to $-\infty$. The masked Q-value matrix is then flattened into a N_{branch}^2-dimensional vector.

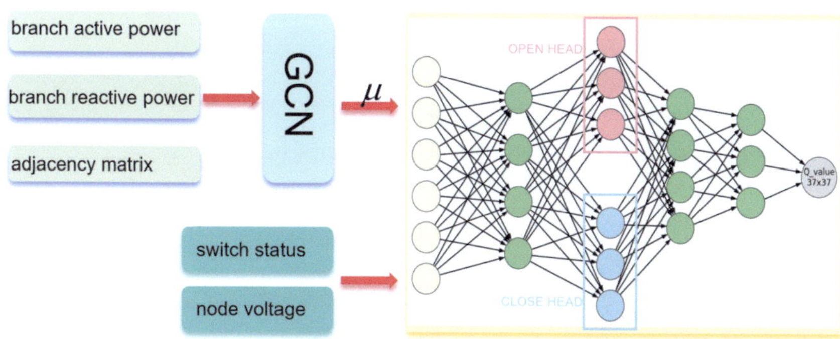

Fig. 2 The architecture of the proposed GCN-DQN network

While the standard DQN algorithm is known to suffer from overoptimistic value estimates due to max operator, this paper employs the Double DQN update mechanism to mitigate this issue. The Double DQN modifies this target by using the current online network Q to select the best action for the next state s', while the target network $\hat{Q}$ is used to evaluate the Q-value of that action:$y = r + \gamma\hat{Q}\left(s', \arg\max Q\left(s', a'\right)\right)$. This decoupling effectively reduces the overestimation of Q-values, leading to more stable and reliable training. The corresponding loss function for our GCN-based Q-network is: $L(\theta) = E_{(s,\, a,\, r,\, s')\sim D\left[(y-Q(s,\, a;\theta))^2\right]}$, where θ represents the parameters of the online Q-network, and D is the experience replay buffer from which transitions $\left(s, a,\ r,\ s'\right)$ are sampled.

3.4 *Masking Mechanism*

Basic Action Mask. The basic action mask $M_{basic} \in R^{N_{branch}\times N_{branch}}$ is generated based on the switch status vector S_{switch}, aiming to prevent meaningless actions. Its mathematical expression is given by Eq. (11):

$$M_{action}\left[i,\ j\right] = \begin{cases} 0, & if\ S[i] = 0\ and\ S[j] = 1\ and\ i \neq j \\ -\infty, & otherwise \end{cases} \tag{11}$$

Topological Feasibility Mask. The basic action mask ensures validity at the switch status level but cannot guarantee topological feasibility. This section designs a mask based on loop detection.

The core principle of this mechanism is that a feasible reconfiguration action must satisfy the condition that after closing a branch, a unique loop is formed, and the branch to be opened b_{open} is an originally closed branch within this loop. Opening branch b_{open} restores the network to a feasible radial structure.

3.5 *Training Process*

The overall training process is outlined in Algorithm 1 below.

Algorithm 1: GCN-DQN Training Algorithm

input: Power data and topology information
output: network weight
Initialize the current Q-network Q_{θ} and the target Q-network $Q_{\theta'}$ with the same parameters.
Initialize the experience replay buffer D with capacity N.
for episode = 1 to M **do**
  Initialize the environment state s_0.
    for t = 1 to T **do**
    With probability ε select a random action a_t.
    Otherwise, select $a_t = \arg\max Q(s_t, a; \theta)$
    Execute action a_t in the environment, observe reward r_t, and next state s_{t+1}.
    Store the experience tuple $\{s_t, a_t, r_t, s_{t+1}\}$ in buffer D.
    Sample a random batch of experience tuples from D.
    for each sampled tuple **do**
     if s_{t+1} is a terminal state **then**
       Set $y_j = r_j$
      else
       Set $y_j = r_j + \gamma * \max_{a'} \hat{Q}(s_{j+1}, \text{argmax} Q(s_{j+1}, a; \theta); \theta')$
      end if
    end for
    Perform a gradient descent step on the loss function
$L(\theta) = (y_j - Q(s_j, a_j; \theta))^2$ to update Q.
    Every C steps, update the target network: $\hat{Q} \longleftarrow Q$.
  end for
end for

4 Performance Evaluation

4.1 Experimental Setup

To validate the effectiveness of the proposed GCN-DQN method, simulation experiments were conducted on the IEEE 33-node distribution system. Its performance was compared against a Genetic Algorithm (GA) and several other DRL variants.

Test System: The IEEE 33-node system was employed, comprising 33 buses and 37 branches (including 5 tie-lines), with a total load of 3.715 MW + j2.300 Mvar. To simulate a practical environment, 20% Gaussian noise was introduced into branch power. Compared Algorithms:

Genetic Algorithm (GA). Chosen as the traditional optimization benchmark, this algorithm is provided with perfect model information and noise-free input data, representing the theoretical performance upper bound under ideal conditions. Parameters are set as follows: population size 50, maximum generations 100, using tournament selection and single-point crossover.

GCN-DQN (standard DQN). This variant uses the GCN architecture but employs the standard DQN update rule. The comparison between this and the proposed method isolates the contribution of the Double-DQN mechanism.

Double-DQN (MLP w/o GCN). This variant replaced the GCN module with a standard Multi-Layer Perceptron (MLP) to process node features, isolating the contribution of the graph convolutional layers in capturing topological information.

GCN-DQN (w/o Encouragement). This was an ablated version of our proposed method where the improvement reward term $\beta\left(P_{loss}^{old} - P_{loss}^{new}\right) / P_{loss}^{old}$ in the reward function was removed. This tests the necessity of the designed nonlinear reward factor.

GCN-DQN Hyperparameters. Learning rate $\alpha = 0.0001$, discount factor $\gamma = 0.9$, experience replay buffer capacity N = 12,000, batch size B = 64, target network update frequency C = 30 steps, exploration rate ε linearly decayed from 1.0 to 0.1.

4.2 Results and Analysis

This section presents a comparative analysis based on a set of 50 test cases with different initial topologies, distinct from the training set. All tests were conducted in an environment with 20% measurement noise. The average initial network loss for all topologies is 229 kW.

The statistics of key performance metrics are summarized in Table 1. The decision-making process of different algorithms for the same test sample are showed in Table 2.

Overall Performance Comparison. The proposed GCN-Double-DQN method demonstrates effective optimization capability, reducing the average power loss from 229 to 146 kW, achieving a 36.2% improvement. While there is a 4.3% performance gap compared to the Genetic Algorithm (140 kW), our method offers a decisive advantage in computational efficiency, with decision times under 0.1 s compared to 42 s for GA, making it over 400 times faster and suitable for real-time applications.

Effectiveness of GCN Architecture. The critical role of graph convolutional networks is evident when comparing GCN-based methods with MLP-based approaches. The proposed GCN-Double-DQN (146 kW) outperforms Double-DQN (MLP) (154 kW) by 5.4%, demonstrating that explicitly modeling grid topology through GCN is substantially more effective than processing flattened features through MLPs. This performance gap highlights GCN's capability to capture spatial relationships and topological constraints inherent in distribution networks.

Table 1 Performance metrics comparison

Performance metric	Comparison of various algorithms*				
Average power loss (kW)	146	140	163	154	205
Average optimization margin	36.2%	38.8%	28.8%	32.7%	10.5%
Average decision steps	5.46	–	4.98	5.34	2.92*
Average computation time	<0.1 s	~42 s	<0.1 s	<0.1 s	<0.1 s
Voltage over limit ratio	2%	0%	2%	6.5%	6.9%
Success rate*	100%	100%	68%	92%	20%

* From left to right, they are GCN-Double DQN (proposed), Genetic Algorithm, GCN-DQN (standard DQN), Double-DQN (MLP w/o GCN), GCN-DQN (w/o Encouragement)
* The lower step count for 'GCN-DQN (w/o Encouragement)' is due to premature termination caused by invalid actions or lack of improvement
* The Success Rate is defined as the percentage of test cases in which the algorithm reduces the power loss below 170 kW

Table 2 The decision-making process of different algorithms

Actions in each step	Each step action of each algorithm* [close,open]			
Step 1	[31,35]	[10,6]	[31,16]	[10,6]
Step 2	[4,27]	[6,27]	[16,15]	[31,35]
Step 3	[33,13]	–	–	[33,13]
Step 4	[32,6]	–	–	[4,27]
Power loss (kW)	143.1	193.3	242.6	157.5

* From left to right, they are GCN-Double DQN (proposed), GCN-DQN (standard DQN), GCN-DoubleDQN (w/o Encouragement), Double-DQN (w/o GCN)

Impact of Advanced Q-Learning. The benefits of Double-DQN are confirmed through controlled comparisons. Within the GCN architecture, GCN-Double-DQN (146 kW) shows consistent improvement over GCN-DQN (Standard) (163 kW), indicating that the Double-DQN mechanism effectively mitigates Q-value overestimation bias and enhances learning stability.

Critical Role of Nonlinear Reward Design. The ablation study reveals the indispensable nature of the proposed nonlinear reward function. Removing the improvement reward term causes catastrophic performance degradation, with GCN-Double-DQN (w/o Encouragement) achieving only 10.5% improvement (205 kW) compared to 36.2% with the complete reward function. The success rate plummets from 100 to 20%, indicating that the agent fails to learn effective optimization strategies without the dense guidance provided by the nonlinear reward factor. This demonstrates that the reward function design is crucial for stable learning and effective policy convergence.

Analysis of Decision-making Process (Table 2). A detailed examination of the sequential switch operations in Table 2 provides further insight into the algorithms' search strategies and convergence behaviors. The proposed GCN-Double-DQN method exhibits a stable and purposeful search trajectory. It methodically adjusts the open switches over four steps, ultimately converging to a low-loss configuration (143.1 kW). This demonstrates effective, multi-step planning. In contrast, the GCN-DQN (Standard) converges prematurely after only two steps, settling at a suboptimal loss of 193.3 kW, which aligns with its higher average loss in Table 1 and suggests instability due to Q-value overestimation. The GCN-Double-DQN (w/o Encouragement) fails to make meaningful progress, stagnating after two steps with the highest final loss (242.6 kW) among the DRL methods. This observation directly supports the findings in previous viewpoint, showing that without the improvement reward, the agent lacks the incentive to explore beneficial actions and often terminates early. Finally, the Double-DQN (w/o GCN) also completes a four-step sequence but achieves a higher final loss (157.5 kW) than the proposed method. This indicates that while it can execute a multi-step plan, its MLP-based state representation is less effective at evaluating network states, leading to a inferior final configuration compared to the GCN-equipped agent.

5 Conclusion

This paper proposed a method for distribution network dynamic reconfiguration in noisy environments based on Graph Convolutional Networks and Deep Q-Learning. The innovations of the proposed method lie in the following two aspects: firstly, leveraging GCN's capability to handle graph-structured data to learn the grid's topological and electrical features, enabling rapid decision-making by the agent; secondly, designing a reward function incorporating a nonlinear reward factor, calculated based on the current network loss, which allows the reward function to provide stable and appropriate reward signals across different scenarios. Experimental results demonstrate that while the solution quality of the proposed method shows only a 4.3% gap compared to the Genetic Algorithm, its fast response capability and adaptability to diverse scenarios surpass those of traditional methods. Future research will focus on the following directions:

Investigating the transfer and generalization capabilities of the model and its training mechanism to large-scale distribution networks.

Further studying the relationship between reward function design and the learning stability and convergence speed of the agent.

Researching multi-timescale dynamic reconfiguration strategies, incorporating distributed energy resources such as distributed photovoltaics into the state space and reward function design.

References

1. Graine et al (2023) A mixed integer linear programming model for the distribution network reconfiguration problem. In: 2023 IEEE 11th international conference on systems and control (ICSC). IEEE, Sousse, Tunisia, pp 70–75
2. Quang PP et al (2023) Distribution network reconfiguration using genetic algorithm considering load profile and the penetration of distributed generation. In: 2023 Asia meeting on environment and electrical engineering (EEE-AM). IEEE, Hanoi, Vietnam, pp 01–06
3. Kahouli O et al (2022) Distribution network reconfiguration for reliability enhancement via genetic algorithm approach. In: 2022 5th International conference on power electronics and their applications (ICPEA). IEEE, Hail, Saudi Arabia, pp 1–6
4. Li T et al (2024) Distribution network reconfiguration based on improved particle swarm optimization algorithm. In: 2024 3rd International conference on energy, power and electrical technology (ICEPET). IEEE, Chengdu, China, pp 746–749
5. Wu X et al (2023) Distribution network reconfiguration based on improved particle swarm optimization algorithm. In: 2023 IEEE 7th information technology and mechatronics engineering conference (ITOEC). IEEE, Chongqing, China, pp 971–975
6. Wang B et al (2022) Application of ant colony algorithm in reconfiguration of power system distribution network. In: 2022 4th International conference on intelligent information processing (IIP). IEEE, Guangzhou, China, pp 241–244
7. Cao D et al Deep reinforcement learning in power systems resilience: a review. IEEE Trans Reliab. https://doi.org/10.1109/TR.2025.3606215
8. Cao D et al (2020) Reinforcement learning and its applications in modern power and energy systems: a review. J Mod Power Syst Clean Energy 8(6):1029–1042
9. Cao D et al (2024) Physics-informed graphical representation-enabled deep reinforcement learning for robust distribution system voltage control. IEEE Trans Smart Grid 15(1):233–246

Droop Control for DC Multi-microgrid Systems Based on Consensus Algorithm

Mingchao Zhang, Wei Pang, Jingbin Yang, Xinrui Liu, Yubo Zhao, Xing Wei, Haoli Liu, and Chenrui Hao

Abstract In distributed DC multi-microgrid systems with energy storage batteries, Each sub-microgrid achieve coordinated power distribution according to local operating conditions by droop control while maintaining DC bus voltage. This enhances power supply quality and improves system reliability. Based on traditional droop control for DC multi-microgrids, an improved droop control strategy is proposes that incorporates a consensus algorithm in this paper. This strategy allows each sub-microgrid to obtain the State of charge (SOC) of multi-microgrid battery clusters and adjust local battery operation modes based on both the system-wide and local battery SOC information. Under this control strategy, the batteries has faster convergence speed and longer operating life. Finally, simulation verification shows that the proposed method effectively addresses bus voltage stability and battery SOC convergence rate issues in multi-microgrid systems.

Keywords Multi-microgrid · Droop control · Consensus algorithm · State of charge (SOC)

M. Zhang (✉) · W. Pang · J. Yang · Y. Zhao · X. Wei · H. Liu · C. Hao
Chaoyang Power Supply Company, State Grid Liaoning Electric Power Co., Ltd., Chaoyang, China
e-mail: 2445263368@qq.com

H. Liu
e-mail: haimingcyj@163.com

C. Hao
e-mail: 1148880688@qq.com

X. Liu
School of Information Science and Engineering, Northeastern University, Shenyang, China
e-mail: liuxinrui@ise.neu.edu.cn

G. Mu et al. (eds.), *Proceedings of the 2nd Electrical Artificial Intelligence Conference, Volume 1*, Lecture Notes in Electrical Engineering 1587,
https://doi.org/10.1007/978-981-95-7932-7_3

1 Introduction

In recent years, coordinated interconnection and power mutual aid among multiple microgrids have become a focal research topic in the field of microgrid systems. Multiple microgrids with parallel and series connections is defined as the multi-microgrid system [1]. Under given scheduling and control systems, the collaborative operation of multiple microgrids enables orderly energy flow between sub-microgrids, significantly reducing energy storage system requirements compared to single microgrids, while substantially improving the stability, economic efficiency, and power supply quality of distributed generation systems. With the rapid development of wind power technology in recent years, the scheduling and control of multi-microgrid systems have become considerably more complex than those of single microgrids. These systems must not only ensure stable operation of internal microgrid components but also achieve coordinated control among individual microgrids to maintain stable operation of the entire multi-microgrid system. The centralized control of multi-microgrid systems features fast response times and high control accuracy, but it imposes higher demands on communication systems. In contrast, communication-free distributed control in multi-microgrid systems has slower response speeds and cannot meet the requirements of systems demanding higher control precision.

With the widespread adoption of distributed control architectures in smart grids [2, 3], Multi-agent consensus algorithms have emerged as a novel approach to addressing operational control challenges in multi-microgrid systems. These consensus algorithms have been extensively applied in power system dispatch and distributed control research [4–6]. Compared to centralized control, hierarchical control based on consensus algorithms significantly reduces the dependence on communication infrastructure [7]. Furthermore, unlike communication-free distributed control methods, consensus-based hierarchical control can meet the stringent precision requirements of complex multi-microgrid systems.

In distributed DC multi-microgrid systems with battery energy storage, droop control enables each sub-microgrid to coordinate power distribution based on local operating conditions while maintaining DC bus voltage, thereby enhancing power supply quality and improving system reliability [8]. However, there are certain limitations in traditional droop control:

(1) Fixed droop coefficients lead to poor flexibility.
(2) It fails to ensure precise DC bus voltage regulation across microgrids.
(3) A trade-off exists between current (power) sharing accuracy and voltage deviation [9].
(4) Controlled units operate independently with limited operational modes.

A state-of-charge (SOC)-based adaptive droop control method is proposed in paper [10] that successfully addresses Limitation (1) (fixed droop coefficients), but remains ineffective against Limitations (2)–(4). Paper [11] further proposed that a variable sag coefficient adaptive sag control can ensure that the DC bus voltage fluctuates within a certain range, but it cannot meet the requirements of higher precision, so that the DC bus voltage can be stabilized at the voltage reference value. A voltage-feedforward adaptive droop control strategy is introduced in paper [12] that effectively resolves Limitations (1)–(3). However, this method still fails to overcome Limitation (4) regarding cooperative unit operation.

The above analysis demonstrates that in both conventional and improved droop control schemes, battery units operate independently without access to the overall state-of-charge (SOC) information of the multi-microgrid battery cluster, preventing SOC-based power output adjustment. To address this limitation, this paper proposes a hybrid control strategy that integrates consensus algorithms with droop control. The proposed approach enables:

(1) Coordinated SOC Management: Individual batteries dynamically regulate their operating states according to the cluster's real-time SOC status;
(2) Dual Performance Optimization: Simultaneously maintains DC bus voltage accuracy while extending battery service life through balanced energy distribution.

2 DC Multi-microgrid Systems

A system composed of geographically adjacent DC microgrids operating in cluster form is defined as a DC multi-microgrid. In DC multi-microgrid systems, the DC bus voltage between microgrids serves as a critical indicator for determining normal system operation and forms the core focus for subsequent discussions on multi-microgrid control strategies. A DC multi-microgrid is a system consisting of interconnected, distributed medium-voltage or low-voltage DC microgrids with specific functions and operational objectives. Within the multi-microgrid system, sub-microgrid must not only operate independently as individual microgrids but also accept and execute cluster-level dispatching commands to achieve system-wide operational goals. The schematic diagram of a DC multi-microgrid structure is shown in Fig. 1.

3 System Mathematical Model

3.1 Mathematical Model of Droop Control

A control method that introduces current feedback and virtual impedance for bus voltage regulation in DC multi-microgrid systems is refered in U-I droop control, enabling dynamic correction of the given reference voltage value. The control structure features a droop curve as the outermost loop in a dual-loop control architecture

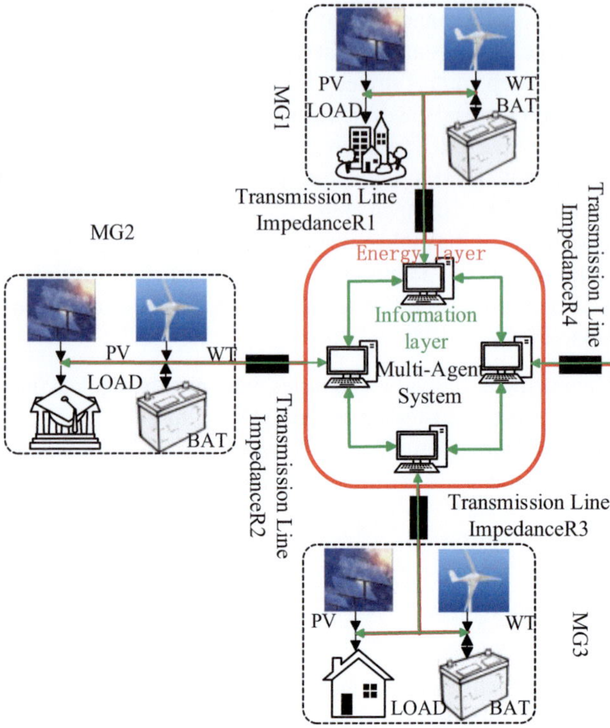

Fig. 1 Schematic diagram of multi-microgrid framework and multi-agent communication

consisting of an outer voltage loop and an inner current loop. The corresponding droop control characteristic curve is shown in Fig. 2.

The droop characteristic expression for DC multi-microgrid systems is given by Wang and Ran [13]:

$$\begin{cases} U_{dc1} = U_{dc1}^{*} - R_1 I_{dc1} \\ U_{dc2} = U_{dc2}^{*} - R_2 I_{dc2} \end{cases} \tag{1}$$

In the equation: U_{dc1} and U_{dc2} represent the output voltages of the two DC/DC converters after droop control; U_{dc1}^{*} and U_{dc2}^{*} denote the reference output voltages of the two DC-DC converters under no-load conditions; R_1 and R_2 are the droop coefficients of the two DC/DC converters, respectively; I_{dc1} and I_{dc2} are the actual output currents of the two DC/DC converters, respectively.

The equivalent circuit diagram of two energy storage batteries with equal sub-microgrid capacities connected in parallel to the DC bus of a microgrid is shown in Fig. 3.

In the figure, r_1 and r_2 represent the transmission impedances from the two converters to the DC bus, respectively, while I_{LOAD} and R_{LOAD} denote the input

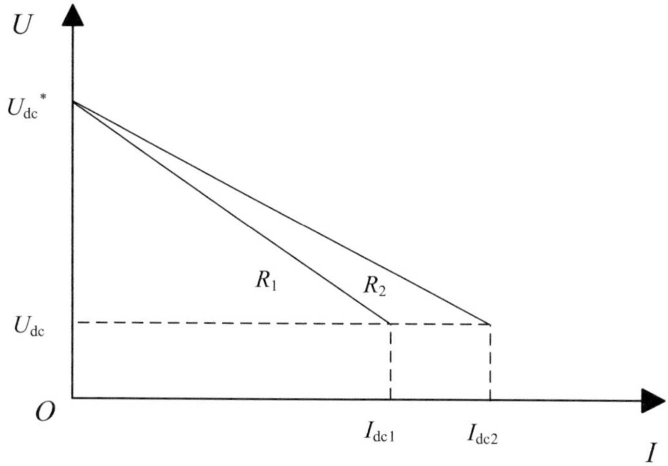

Fig. 2 Droop control characteristic curve

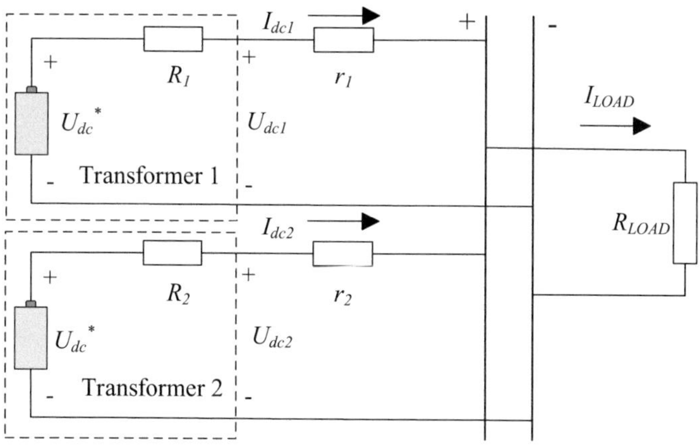

Fig. 3 Equivalent circuit diagram of two converters in parallel

current and equivalent impedance of the load on the DC bus. The KVL and KCL equations for the two transformer columns in Fig. 3 are shown below:

$$U_{\mathrm{dc}}^{*} = I_{\mathrm{dc1}}(R_1 + r_1) + I_{\mathrm{LOAD}}R_{\mathrm{LOAD}} \tag{2}$$

$$U_{\mathrm{dc}}^{*} = I_{\mathrm{dc2}}(R_2 + r_2) + I_{\mathrm{LOAD}}R_{\mathrm{LOAD}} \tag{3}$$

$$I_{\mathrm{LOAD}} = I_{\mathrm{dc1}} + I_{\mathrm{dc2}} \tag{4}$$

From Eqs. (2)–(4), the ratio of the output currents I_{dc1} and I_{dc2} of the two converters can be derived as follows:

$$\frac{I_{dc1}}{I_{dc2}} = \frac{R_2 + r_2}{R_1 + r_1} \tag{5}$$

Equation (5) indicates that the actual output current of a DC/DC converter is influenced by two parameters: the droop coefficient and the transmission impedance. When the impact of transmission impedance is neglected, the relationship simplifies to:

$$\begin{cases} R_1 \gg r_1 \\ R_2 \gg r_2 \end{cases} \tag{6}$$

3.2 Consensus Algorithm Mathematical Model

The first-order multi-agent consistency algorithm is simple in form and has a wide range of applications, and the dynamic expression of the first-order consistency algorithm is as follows:

$$\dot{x}_i(t) = u_i(t), \quad i = 1, 2, 3 \ldots, N \tag{7}$$

In the equation: $x_i \in R^{\mathrm{m}}$ and $u_i \in R^{\mathrm{n}}$ represent the state variable and control input variable of agent i, respectively. The first-order multi-agent system consensus protocol [14] is given as follows:

$$u_i(t) = -\sum_{j=1}^{N} a_{ij}(x_i - x_j), i = 1, 2, 3, \ldots, N \tag{8}$$

Introducing the Laplace matrix and bringing (8) into (7), the resulting linear system is as follows:

$$\dot{x}(t) = -(L \otimes I_m)x(t) \tag{9}$$

In the equation:

$$x(t) = \left(x_1^T(t), x_2^T(t), x_3^T(t), \ldots, x_N^T(t)\right)^T \in R^{Nm} \tag{10}$$

Under the action of controller (8), the system can achieve consistency, and the final convergence value is:

$$x_i \to x_j \to \sum_{i=1}^{N} \xi_i x_i(0) \tag{11}$$

In the equation: $\xi = [\xi_1, \xi_2, \xi_3, \ldots, \xi_N]^T$, $\xi_i \geq 0$ is the left eigenvector corresponding to the eigenvalue 0 of the Laplacian matrix L and satisfies $1_N^T \xi = 0$. The above analysis shows that the final consensus value depends on both the agents' initial states and the communication topology.

Paper [15] investigated the consensus problem for discrete-time multi-agent systems that is shown in Eq. (12) and established that the system can achieve asymptotic average consensus when the network topology is strongly connected and balanced, and $\zeta \in (0, 1/\Delta)$, where Δ represents the maximum degree of network nodes.

$$x_i(t+1) = x_i(t) + \zeta \sum_{j \in N_i} a_{ij}(x_j(t) - x_i(t)),\ i \in N \tag{12}$$

4 Consensus-Algorithm-Based Droop Control for DC Multi-microgrids

This paper integrates consensus algorithms with droop control, enabling each battery unit to determine its operating mode based on both its individual status and the collective state of all batteries. This approach accelerates the SOC convergence of battery clusters, extends battery service life, and consequently reduces costs. The operating principle of the consensus-algorithm-based adaptive droop control is illustrated in Fig. 4.

The working principle of battery state evaluation based on consensus algorithm is presented as follows:

$$SOC_i(k+1) = \sum_{j=1}^{n} d_{ij} SOC_j(k) \tag{13}$$

In the equation: $SOC_i(k)$ represents the state of charge of the ith battery after kth iterations; j denotes the batteries set of communicating with the ith battery; n is the number of batteries connected to the ith battery; d_{ij} represents the state transition coefficient.

To accelerate the convergence rate of battery SOC, the state transition coefficient α is determined according to the Metropolis method [16] as follows:

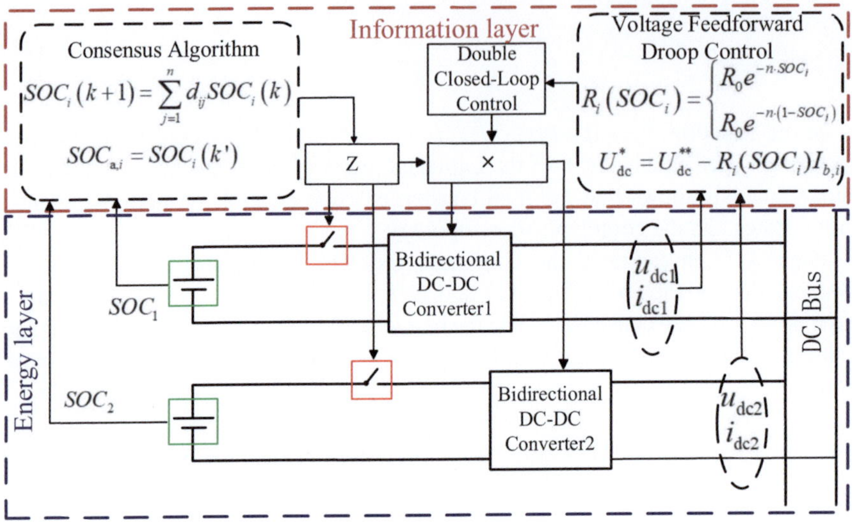

Fig. 4 Schematic diagram of adaptive droop control based on consensus algorithm

$$d_{ij} = \begin{cases} 1/(n_i + n_j + 1), \ j \in N_i \\ 1 - \sum\limits_{j \in N_i} 1/(n_i + n_j + 1), \ i = j \\ 0, \ \text{others} \end{cases} \tag{14}$$

In the equation: N_i represents the adjacency matrix of ith agent; n_i and n_j denote the degrees (number of connected batteries) of the ith and jth battery units, respectively.

Each battery unit can obtain the global average SOC of the battery cluster as follows:

$$SOC_{a,i} = SOC_i(k') \tag{15}$$

In the equation: $SOC_{a,i}$ represents the average SOC of the battery cluster obtained by the ith battery unit; $SOC_i(k')$ denotes the SOC of the ith battery unit after k'th iterations. After sufficient iterations, the SOC average obtained by each battery unit satisfies:

$$\begin{cases} SOC_{a,i} \approx SOC_{a,j} \\ SOC_i(k') - SOC_i(k' - 1) < \xi \end{cases} \tag{16}$$

In the equation: $SOC_{a,j}$ represents the average SOC value obtained by jth battery unit in the cluster; ξ denotes the critical value for the SOC difference between consecutive iterations. The iteration process terminates when the SOC difference becomes smaller than ξ.

After the first iteration round, we can obtain the initial determination of battery operating states as follows:

$$z_i^1 = \begin{cases} 1, \text{ if } I_{\text{b},i} > 0, SOC_i > SOC_{\text{a},i} \\ 0, \text{ if } I_{\text{b},i} > 0, SOC_i < SOC_{\text{a},i} \\ 1, \text{ if } I_{\text{b},i} < 0, SOC_i < SOC_{\text{a},i} \\ 0, \text{ if } I_{\text{b},i} < 0, SOC_i > SOC_{\text{a},i} \end{cases} \tag{17}$$

In the equation: z_i^1 represents the first-round operating state determination of the ith battery unit, $z_i^1 = 1$ indicates participation in droop control $z_i^1 = 0$ signifies withdrawal from operation.

Furthermore, the number of battery units participating in operation after the first-round determination can be obtained as follows:

$$Z^1 = \sum_{i=1}^{n} z_i^1 \tag{18}$$

To ensure stable operation of the multi-microgrid system with sufficient battery units, we define the minimum number of operating batteries as Z_{min}. The operational rules are as follows: If the current number of operating batteries $Z^1 \geq Z_{\text{min}}$, Batteries with working $z_i^1 = 1$ from the first iteration remain operational the other batteries withdraw from operation. If the current number of operating batteries $Z^1 < Z_{\text{min}}$, Batteries with working $z_i^1 = 1$ maintain their operational status, Batteries with working $z_i^1 = 0$ require re-evaluation of their working status. To guarantee SOC convergence while keeping the communication topology unchanged. Batteries with working $z_i^1 = 1$ from previous evaluation continue participating in the new iteration, with their SOC values reset to 0 for the consensus process. Batteries with working $z_i^1 = 0$ maintain their original SOC values during iteration the evaluation process terminates when the number of operating batteries satisfies Eq. (19). The complete battery status determination algorithm based on the consensus algorithm is presented in Fig. 5.

$$Z^m = \sum_{i=1}^{n} z_i^m \geq Z_{\text{min}} \tag{19}$$

In the equation: m represents the number of maximum iteration round; n denotes the total number of battery units in the cluster.

We select 8 batteries with different SOCs, and give the number of execution rounds m of the algorithm when the minimum number of batteries in the discharge and charging states is shown in Fig. 6a, b, respectively.

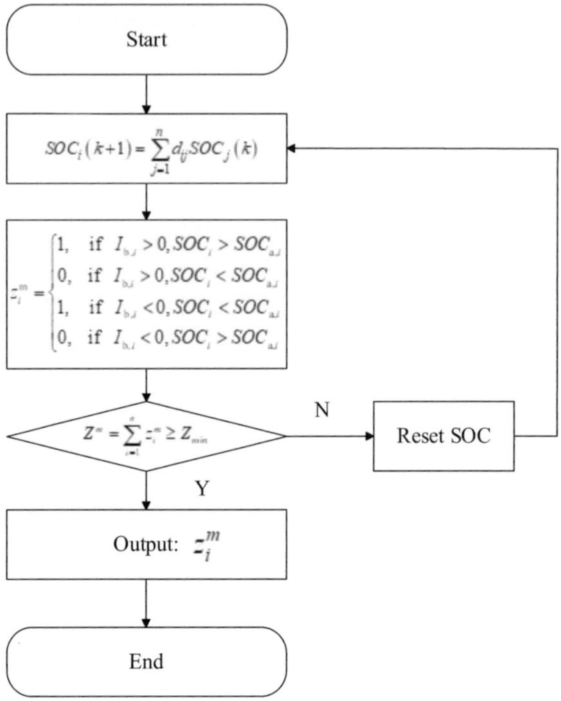

Fig. 5 Algorithm for determining the working state of the battery

5 Simulation Studies

This section simulates and analyzes the operation conditions of multi-microgrid systems under different drooping control strategies, and the initial simulation parameters of multi-microgrid systems under source-load power balance are shown in Table 1.

The sub-microgrid of the multi-microgrid system are balanced in 0–5 s of internal source charge, and the multi-microgrid system is stable after 0.1 s. We set the load of 1th sub-microgrid to increase by 100% within 0.5–1 s, the power of all distributed power sources of 2nd sub-microgrid to fail within 1.5–2 s becomes 0, and the power consumption of 3rd sub-microgrid becomes 0 within 2.5–3 s.

The simulation results under different control strategies are shown in Figs. 7, 8, 9, and 10. In Figs. 7, 8, 9, and 10, (a) to (d) are the changes of state quantity under SOC-based adaptive sag control, variable sag coefficient adaptive droop control, voltage feed-forward adaptive sag control, and the consistency algorithm-based sag control proposed in this paper. The bus voltage change of the multi-microgrid system and the bus voltage change of each subi-microgrid are shown in Fig. 7(1) and (2), respectively. We can see that the bus voltage of the multi-microgrid is stable at its reference value under the control strategy of method (c) and method (d). Figure 8(1) and (2) show the battery current (power) of each sub-microgrid and the net exchange

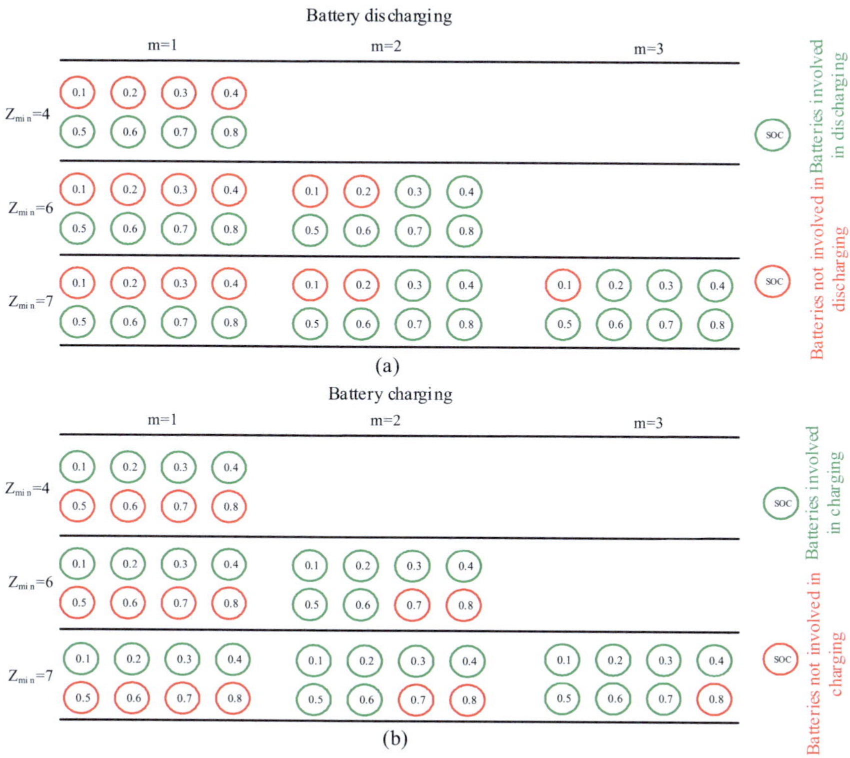

Fig. 6 Schematic diagram of battery operation status

Table 1 Initial values of multi-microgrids simulation parameters

Parameter	Value	Parameter	Value
SOC_1	90/%	SOC_2	80/%
SOC_3	70/%	SOC_4	60/%
P_{LOAD1}	4000/W	P_{LOAD2}	4000/W
P_{LOAD3}	4000/W	P_{LOAD4}	4000/W
P_{DG1}	4000/W	P_{DG2}	4000/W
P_{DG3}	4000/W	P_{DG4}	4000/W
R_1	1	R_2	2
R_3	3	R_4	4
r_1	2.30/Ω	r_2	2.16/Ω
r_3	1.94/Ω	r_4	1.82/Ω

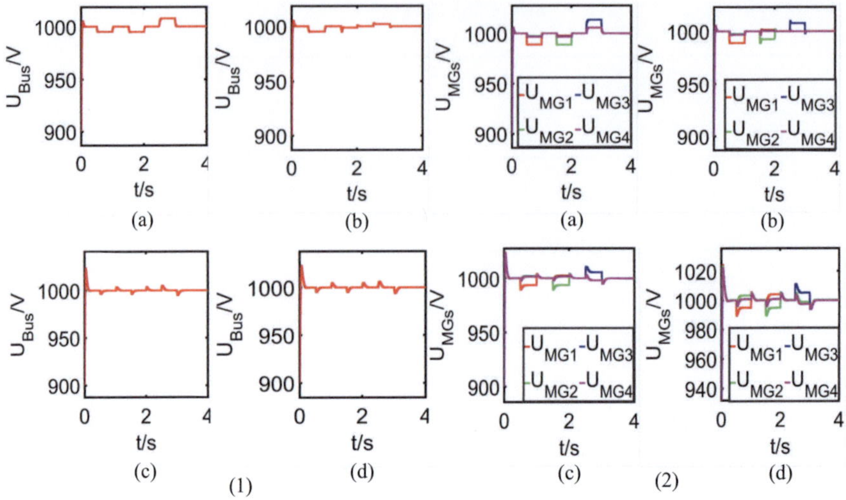

Fig. 7 Bus voltage of multi-microgrid system

current (power) of each sub-microgrid of the battery under the control strategy of method (a) to (d), the discharge current with high SOC is greater when the battery is discharged, and the charging current with low SOC is greater when the battery is charged, and when the SOC is not in the specified value range, it will cause the battery to withdraw from operation. Figure 9 shows the change of battery SOC in each sub-microgrid, and we can see that the SOC converges faster under the control of the control strategy (d) proposed in this paper. The working state of each battery under the sag control strategy based on the consistency algorithm proposed in this paper is shown in Fig. 10.

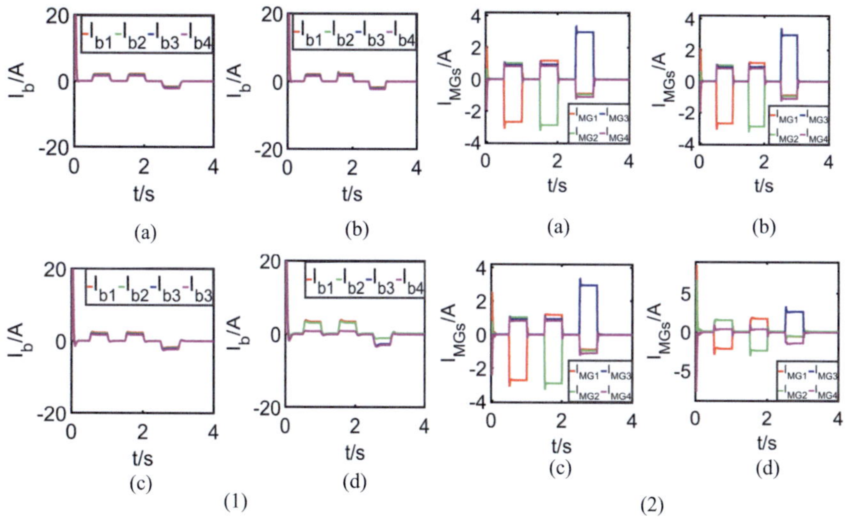

Fig. 8 The battery current of multi-microgrid system

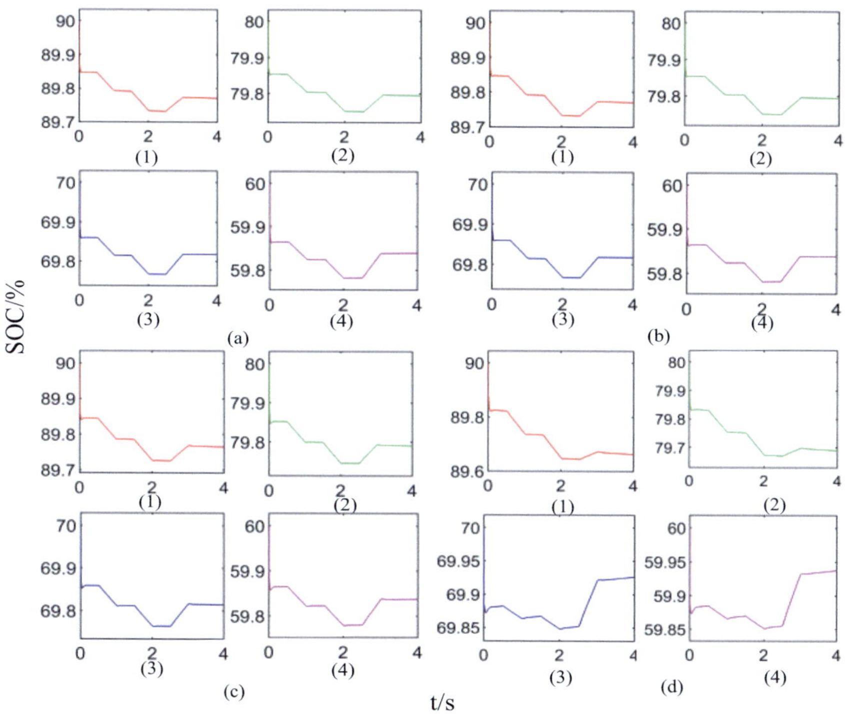

Fig. 9 The SOC of batteries in sub-microgrid

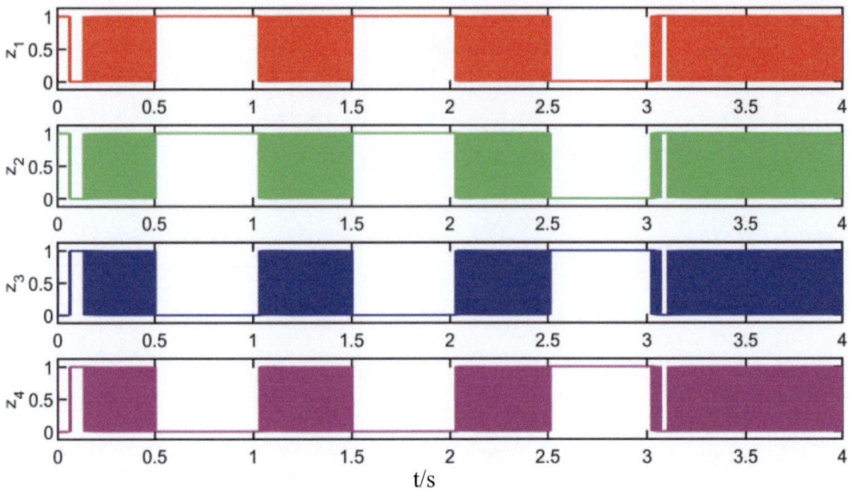

Fig. 10 The operating status of batteries in sub-microgrid

6 Conclusion

In this paper, a droop control strategy based on consistency algorithm is proposed for multi-microgrid systems. The top-level control of this control strategy uses the consistency algorithm to interact information with each subi-microgrid SOC to evaluate the state of the battery pack, and the bottom-level control agents adjust the battery power and other controllable unit operating states according to the multi-microgrid battery SOC and the battery SOC in the subi-microgrid. The control strategy uses a distributed network to overcome the shortcomings of independent operation of each subi-microgrid, and proposes to judge the conditions for battery participation and battery power regulation to accelerate the SOC convergence speed of the battery pack and extend the service life of the battery pack while keeping the bus voltage of the multi-microgrid system stable.

References

1. Zhao Z, Guo J, Luo X et al (2022) Distributed robust model predictive control-based energy management strategy for islanded multi-microgrids considering uncertainty. IEEE Trans Smart Grid 13(3):2107–2120
2. Rahbari-Asr N, Ojha U, Zhang ZA, Chow MY (2014) Incremental welfare consensus algorithm for cooperative distributed generation/demand response in smart grid. IEEE Trans Smart Grid 5(6):2836–2845
3. Yang G, Qian A, Jing W et al (2014) Consensus cooperative control of AC/DC hybrid microgrids based on multi-agent system. High Volt Eng 5(6):2836–2845

4. Wu KM, Li Q, Chen ZY et al (2021) Distributed optimization method with weighted gradients for economic dispatch problem of multi-microgrid systems. Energy 222
5. Zhang B, Wang J, Liang D et al (2020) Optimization control strategy of distributed energy storage in autonomous microgrid cluster on consensus algorithm. Power Syst Technol 44(5):1705–1713
6. He Y, Fan L, Han B et al (2017) A consensus protocol based control method for coordinated multi-microgrid control. Power Syst Technol 41(4):1269–1276
7. Yazdanian M, Mehrizi-sani A (2014) Distributed control techniques in microgrids. IEEE Trans Smart Grid 5(6):2901–2909
8. Shekhar S, Alam A (2025) A transient-adaptive droop control strategy for better current sharing, robustness, and stability in DC-MGs. IEEE Trans Ind Appl 61(4):5513–5524
9. Zhu S (2019) The research of droop control in DC microgrid. Shanghai University
10. Zhang L, Yan K, Leng X et al (2021) Research on coordinated control strategy of an autonomous DC microgrid based on SOC droop control. Power Syst Protect Control 49(12):87–97
11. Ning X, Pan H, Li F et al (2024) An improved droop control strategy for a DC microgrid based on line resistance observations. Power Syst Protect Control 52(11):42–51
12. Liu H, Liu C, Zheng L et al (2022) Cooperative optimal droop control for VSC-MTDC system with quasi non-error DC voltage regulation. Autom Electr Power Syst 46(6):117–126
13. Wang Q, Ran P (2023) Control strategy of current equalization of energy storage systems and voltage stability of DC bus in isolated microgrid. Proc CSU-EPSA 35(6):132–140
14. He W, Chen Z, Hu F et al (2023) Consensus-based low-carbon economic dispatching of integrated energy systems. Power Syst Protect Control 51(19):42–53
15. Olfati-Saber R, Fax JA, Murray RM (2007) Consensus and cooperation in networked multi-agent systems. Proc IEEE 95(1):215–233
16. Lu M, Lu L (2017) Cooperative output regulation of linear multi-agent systems by a novel distributed dynamic compensator. IEEE Trans Autom Control 62(12):6481–6488

Motor Fault Diagnosis Method Based on Multi-scale Gray Texture Image and Dual Attention Mechanism

Shengshi Yang, Xiaofei Zhang, Guojun Qin, and Fan Hu

Abstract In recent years, deep learning technology has achieved breakthrough development in the field of mechanical and electrical equipment fault diagnosis, but the research on the problem of motor vibration signal fault feature representation and model channel space correlation learning is still insufficient. Therefore, based on the existing research, this study proposes a motor fault diagnosis method based on multi-scale texture enhancement and dual attention mechanism. Firstly, by constructing multi-scale gray texture image (MGTI), the high fidelity conversion of vibration signal from one-dimensional time domain to two-dimensional texture space is realized. Secondly, by introducing the channel space dual attention module (CBAM), an improved residual network (CBRN) is designed to adaptively calibrate the feature response to focus on the key fault information. Finally, the experiments are conducted on the induction motor (IM) show that the accuracy of the eight types of motor fault diagnosis is 99.75%, which is up to 12.25% higher than the traditional fault diagnosis model (such as SVM). The experimental results confirm the feasibility and effectiveness of CBRN in the field of fault diagnosis.

Keywords Convolutional block attention module (CBAM) · Multi-scale gray texture image (MGTI) · Fault diagnosis · Residual network

1 Introduction

As the core power component of modern industrial equipment, the operation state of motor directly affects the safety and reliability of production system [1]. When the motor runs for a long time in harsh environment, it will inevitably fail. If the fault cannot be detected in time, it will lead to huge economic losses and even

S. Yang · X. Zhang (✉) · G. Qin
Hunan University, Changsha, China
e-mail: zhangxiaofei@hnu.edu.cn

F. Hu
National University of Defense Technology, Changsha, China

G. Mu et al. (eds.), *Proceedings of the 2nd Electrical Artificial Intelligence Conference, Volume 1*, Lecture Notes in Electrical Engineering 1587,
https://doi.org/10.1007/978-981-95-7932-7_4

potential safety hazards [2]. However, mechanical and electrical faults such as rotor fracture, bearing wear, winding short circuit and so on are easily caused by the motor under complex working conditions for a long time. The traditional motor fault diagnosis is mainly based on expert knowledge system [3] and manual feature extraction engineering [4], generally have limitations such as insufficient feature extraction efficiency and weak anti-interference ability. In contrast, the deep learning technology represented by neural network has achieved large-scale application in the field of motor fault diagnosis. It has effectively broken through the bottleneck of traditional methods through the data-driven feature learning mechanism and formed research results with reference value [5, 6].

As a classic architecture in the field of deep learning, convolutional neural network (CNN) has become a research hotspot with its excellent feature representation ability. For one-dimensional time-series signal processing, 1DCNN achieves efficient feature extraction through local sensing mechanism. Typical applications such as the hybrid architecture of 1DCNN and multiscale graph convolution constructed by Xiao et al. [7] effectively improve the fault detection accuracy of motor fault diagnosis. Ye et al. [8] verified the robustness of the method in bearing health state identification through the collaborative optimization of variational mode decomposition and 1DCNN. The new 1DCNN variant proposed by Wang et al. [9] achieved a breakthrough in model convergence speed and training stability. In terms of two-dimensional feature analysis, 2DCNN encodes the original signal into image features through time–frequency domain conversion method. For example, Nie et al. [10] used multi-channel vibration signal RGB coding technology to construct multidimensional fault representation. Zhang et al. [11] proposed a motor fault diagnosis framework based on meta learning by fusing multi-sensor images. It is noteworthy that Zhang et al. [12] proposed the multi primitive energy mode (MBEP) conversion method, which sensibly improves the accuracy of motor fault discrimination by suppressing the influence of working condition disturbance on image features. Wang et al. [13] innovatively combined the four quadrant entropy feature with the multi-scale perception network to achieve efficient diagnosis in a few sample scenarios. It is particularly noteworthy that the symmetrical dot pattern (SDP) coding strategy developed by Huang et al. [14] performs well in the small sample diagnosis of local demagnetization fault of the PMSM. Liu et al. [15] performed continuous wavelet transform (CWT) on the signal and proposed a diagnosis framework based on transfer learning, which effectively improved the transfer generalization ability of fault features. These research results systematically verify the dual advantages of CNN architecture in time sequence signal analysis and image feature learning. In addition, transforming the signal into an image to characterize its characteristics is helpful to improve the diagnostic performance. However, the existing methods still face two major challenges:

(1) The vibration signal of the motor surface under complex working conditions presents significant non-stationary characteristics, and the factors such as strong electromagnetic interference, multi-source sensor coupling noise and nonlinear distortion of transmission link in the operating environment of the equipment are superimposed, resulting in serious deterioration of the SNR of the original

data, and the robust characterization of key fault characteristic information is facing severe challenges.

(2) Conventional neural network has insufficient capability to model multi-scale image features and channel-spatial correlation, and ignores the synergy between channel dimension and spatial domain of feature map, which further limits the exaltation of diagnostic accuracy.

To solve the above problems, this paper improved a motor fault diagnosis method based on multi-scale texture image and dual attention mechanism. The main innovations are reflected in three aspects:

(1) A multi-scale grayscale texture image (MGTI) generation method is constructed. The high fidelity transformation of signal features from one-dimensional time domain to two-dimensional texture space is realized by segmenting the vibration signal, multi-scale circular binary pattern (MCLBP) texture enhancement and bicubic interpolation optimization.
(2) An improved residual network (CBRN) based on the CBAM module is designed to dynamically calibrate the feature response through the channel-space dual attention mechanism to enhance the focus ability of key fault features.
(3) This study constructed diagnostic models for eight common motor faults and validated their effectiveness on an independently developed induction motor testing platform. The experimental results demonstrate that even with limited sample size, the model can still achieve an average diagnostic accuracy of up to 99.75%, and its performance improvement can reach up to 12.25% of traditional fault diagnosis models.

2 Related Work

2.1 The Theory of MGTI

The vibration signal is sampled by fixed length window segmentation, and the two-dimensional feature tensor is constructed by structural recombination. The tensor value is normalized to the [0–255] interval by the linear normalization method to realize the accurate mapping of the gray pixel value, and finally generate the gray image representing the signal characteristics [16]. The conversion process follows the following numerical correspondence formula:

$$X_{i,j} = 255 \times \frac{x_{i,j} - Min(x)}{Max(x) - Min(x)} \tag{1}$$

where, i and j are the coordinates values in the feature tensor, $x_{i,j}$ is the original data point value of the corresponding position in the feature tensor, $Max(x)$ is the maximum value and $Min(x)$ is the minimum value, and $X_{i,j}$ is the value of the coordinate position corresponding to the grayscale image.

The multi-scale circular binary pattern (MCLBP) [17] is used to set multiple radius values (such as $r_1, r_2, \ldots, r_n$, and the corresponding weight is $w_1, w_2, \ldots, w_n$). Each radius corresponds to a circular neighborhood of the gray image, in which P sampling points are evenly distributed on the circumference of each radius.

The coordinates of the central point are (x_c, y_c), and the central pixel gray value is g_c. The coordinates of the p sampling point are calculated by formula (2), but because the calculation results may have non integer values, they cannot directly correspond to the discrete pixel units of the image matrix. Therefore, the bilinear interpolation method [18] can be used to accurately estimate the pixel gray value g_p corresponding to the non-integer coordinate position.

$$\begin{cases} x_p = x_c + r \cdot \cos\left(\frac{2\pi p}{P}\right) \\ y_p = y_c + r \cdot \sin\left(\frac{2\pi p}{P}\right) \end{cases} \tag{2}$$

For each radius r, compare the central pixel gray value g_c with the pixel gray values g_p of the surrounding P sampling points. By comparing the size relationship between the gray value of adjacent pixels and the gray value of the center point, when the gray value of peripheral pixels is not lower than the center point, the pixel position is assigned binary 1, and vice versa. According to the sequence of clockwise traversing 8 neighborhood points from the top left corner, the binary sequence obtained is converted into decimal coding, and finally the eigenvalue *LBP* of the central pixel is generated.

$$LBP_{P,r_i} = \sum_{p=0}^{P-1} F\left(g_p - g_c\right) \cdot 2^p \tag{3}$$

$$F(x) = \begin{cases} 1, & x \geq 0 \\ 0, & x \leq 0 \end{cases} \tag{4}$$

The final center value is obtained by weighted summation:

$$LBP = \sum_{p=0}^{P-1} LBP_{P,r_i} \cdot w_i \tag{5}$$

2.2 *Attention Mechanism*

Attention mechanism, as a key technology in deep learning, is designed based on the functional characteristics of the human visual system in filtering information. In the process of visual information processing, human beings can independently choose to focus on different targets or the key areas of the same target. This biological characteristic provides important enlightenment for the design of attention model. According to the differences of application scenarios, attention mechanisms can be divided into two types: channel attention [19] and spatial attention [20].

Channel attention acts on the feature channel dimension, and optimizes the feature learning process by establishing the correlation weight between channels. Its core is to identify the information value differences of different feature channels. The corresponding spatial attention acts on the feature space dimension, and guides model to concentrate on the momentous areas by establishing the spatial position weight. By integrating the above two mechanisms, the hybrid attention mechanism establishes dual attention weights in the spatial dimension and the channel dimension, thereby achieving the comprehensive optimization of feature information and significantly improving the model's ability to represent multi-dimensional features.

3 Proposed Method

Figure 1 clearly illustrates the framework composition of the method proposed in this paper, mainly including MGTI image generation and diagnosis model design. These two parts are described in detail below.

3.1 Signal to MGTI

The vibration signal of the motor surface is collected by the vibration sensor, and then the vibration signal is segmented and sampled by a fixed length window. The two-dimensional feature tensor is constructed by structural reorganization, and the multi-scale gray texture image is constructed by multi-scale circular binary pattern

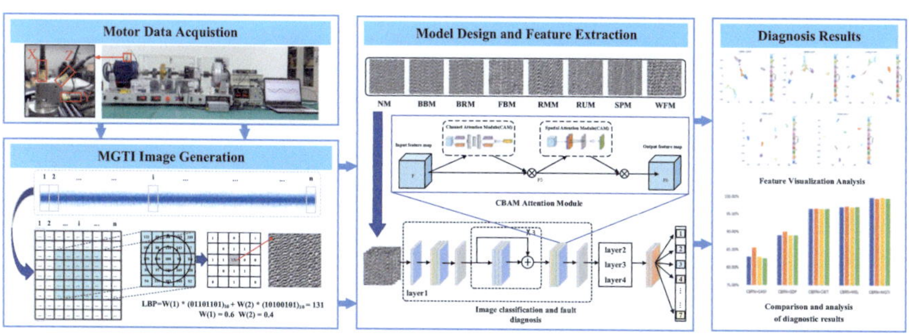

Fig. 1 Framework of the proposed method

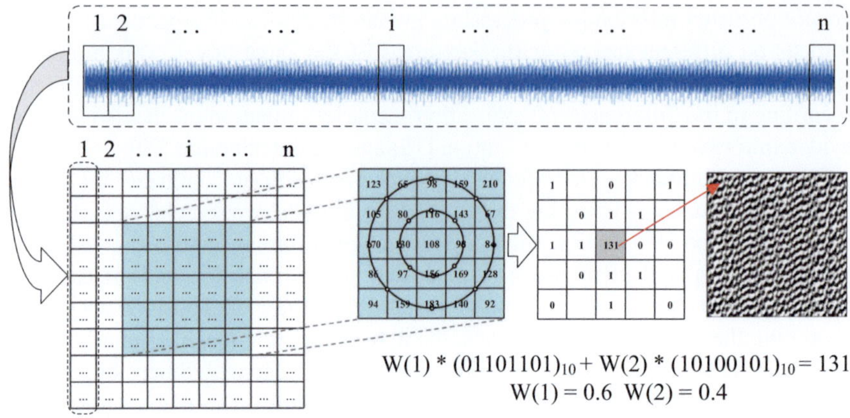

Fig. 2 MGTI conversion process

texture enhancement and bicubic interpolation optimization. The following content is the specific implementation process.

Step 1: the vibration signal is converted to grayscale image. The one-dimensional temporal signal is segmented by using a sliding window with a data length of 64 and reorganized into a two-dimensional feature tensor of 64*64. The matrix elements are linearly normalized by Eq. (1), and the gray-scale quantitative mapping is completed according to the amplitude interval [0–255], and the preliminary vibration characteristic gray-scale image is generated.

Step 2: convert gray image to gray texture image. According to the formula (2–5), two circular binary patterns with different scales are selected for texture feature enhancement. The LBP Operator with 8 sampling points (P) within radius r of 1 and the LBP Operator with 8 sampling points (P) within radius r of 2 are selected. The circular operator expands the neighborhood to 5*5. Combined with the weight distribution, the center image value is calculated.

Step 3: use bicubic interpolation [21] to convert the 64*64 feature matrix into 256*256 to obtain 256*256 multi-scale gray texture image (MGTI). The implementation process is shown in Fig. 2.

3.2 *Dual Attention Mechanism*

By integrating the attention branches with complementary characteristics, the dual attention mechanism can realize the adaptive focus of the model on key features, and show significant performance advantages in complex visual tasks. The core structure of CBAM module adopts the sequential combination of attention units in channel domain and spatial domain: the channel attention module uses global average pooling and maximum pooling to capture features, and generates channel weights by

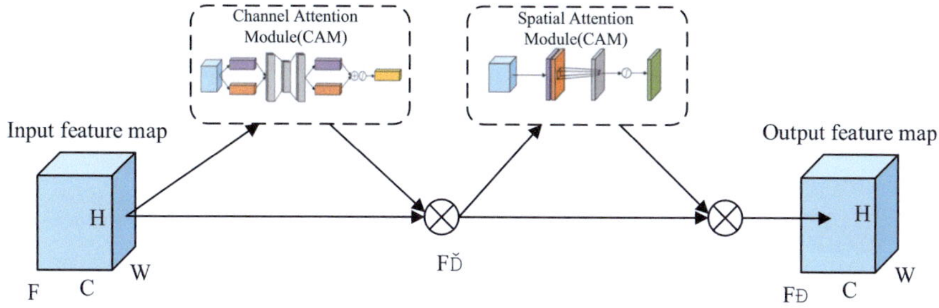

Fig. 3 Structure diagram of CBAM

sharing MLP to strengthen important channels. The spatial attention module pools the compressed features of the channel, and generates spatial weights by 7 × 7 convolution to focus on the key areas.

CBAM module optimizes feature expression by integrating dual attention mechanisms of channel and space, and the structure is shown in Fig. 3. The module adopts a step-by-step processing strategy. Firstly, the adaptive weight allocation is implemented in the channel dimension to focus on the contribution of different characteristic channels; Then dynamic weighting calculation is performed in the spatial dimension to locate the details of key areas. Through this hierarchical feature enhancement method, the model can more efficiently extract cross dimensional salient features, and finally generate an optimized feature map with strong representation ability. This dual attention coordination mechanism effectively improves the capacity of the model to analyze complex features, and enhances the feature selectivity while retaining the original amount of information.

3.3 CBRN Model

ResNet's residual blocks alleviate the gradient vanishing problem by jumping connection, but the convolution layer in each residual block handles all channels and spatial positions indiscriminately, which may lead to the retention of redundant features and insufficient emphasis on key features.

The method proposed in this paper is used to classify the gray texture image so as to classify the motor fault type. The network model CBAM-ResNet is shown in Fig. 4 and named CBRN. The network is improved based on ResNet-50 architecture, which is mainly composed of residual block group, attention module and multi-level pooling layer. The main body of the network consists of four residual block groups (layer1 to layer4), wherein layer1 is stacked with three bottleneck blocks, and layer2, layer3 and layer4 contain four, six and three bottleneck blocks respectively. Each bottleneck block uses a three-layer convolution structure of 1 × 1, 3 × 3 and 1 × 1.

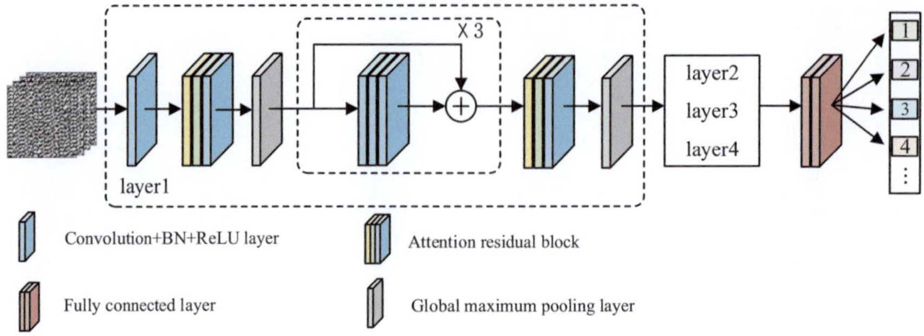

Fig. 4 Structure diagram of CBRN model

The convolution kernel step is set to 1, and the 3 × 3 convolution layer is filled with 1 pixel to maintain the size of the feature map.

Each convolutional layer is succeeded by a batch normalization layer and a ReLU activation function. This design mitigates the gradient vanishing problem and improves feature reusability via residual connections. In addition, after the initial convolution layer and the output of the first three residual block groups, four CBAM attention modules are embedded in turn to dynamically calibrate the feature response through the channel space dual attention mechanism.

The network Model input first extracts the shallow features through 7 × 7 convolution layer. When the features pass through the residual block group, the CBAM module successively performs channel weight calibration and spatial region focusing on the outputs from layer1 to layer3. The channel attention learns the correlation between channels through global pooling and full connectivity layer, and the spatial attention extracts the spatial significance through pooling splicing and convolution. At the end of the network, the features are compressed into 1 × 1 vectors by global maximum pooling, and finally input to the softmax classifier through the full connection layer to map the network output to different categories. Through the collaborative optimization of attention mechanism and deep residual learning, the design significantly improves the discrimination ability of the model for key features.

4 Experiments Analysis

4.1 Experiment Description

Dataset description. In order to verify the engineering applicability of the proposed method, a multi-state detection experimental system for induction motor is developed. The device is composed of a power transmission unit and a data acquisition unit, and its mechanical structure layout is shown in Fig. 5. As shown in Fig. 6, the experimental platform provides eight types of motors in different states, including normal motor (NM), broken rotor bar motor (BBM), bowed rotor motor (BRM), fault bearing motor (FBM), rotor misalignment motor (RMM), rotor unbalance motor (RUM), single phasing open circuit fault motor (SPM), and winding short circuit fault motor (WFM). The three-axis vibration sensor is used to collect the motor vibration signal and transmit it to the signal acquisition equipment. One million data points are collected for each type of fault, and the relevant parameters during motor operation are shown in Table 1.

After the triaxial vibration signal is collected, the uniaxial signal is converted into multi-scale gray texture image (MGTI). To promote the subsequent comparative experimental analysis, the uniaxial signal is also converted into symmetrized dot pattern (SDP), continuous wavelet transform (CWT), Mel (MEL) and Gramian angular summation field (GASF). The images obtained by uniaxial MGTI and other conversion methods in the same state are shown in Fig. 7.

Experimental setup. Taking eight kinds of motor states as an example, the fault occurrence in the actual industrial scene is simulated, and the vibration sensor is used to collect the signals of eight different motor fault types. Through the proposed

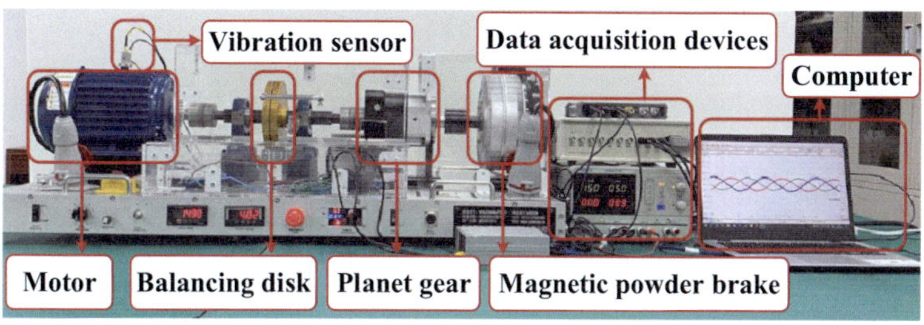

Fig. 5 IM experimental platform

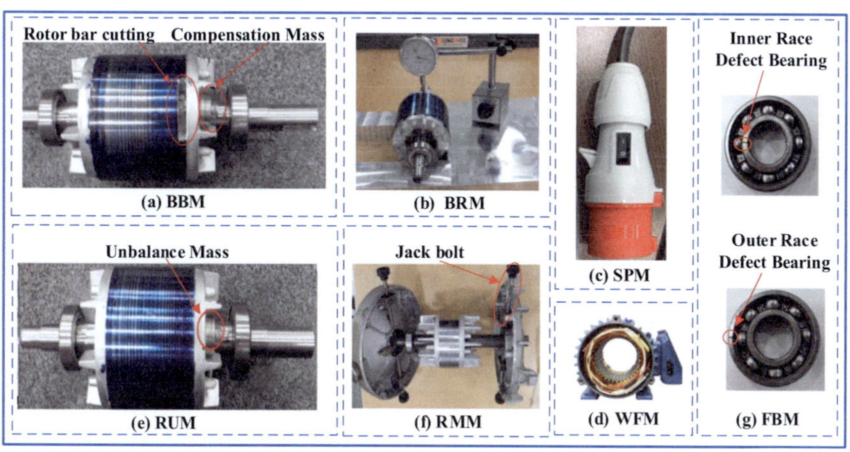

Fig. 6 Motor fault implantation diagram

Table 1 Parameters of motor in operation

Speed	Voltage	Current	Torque	Power	Poles
1000 r/min	220/380 V	6.4/3.4 A	9 N*m	1 hp	4

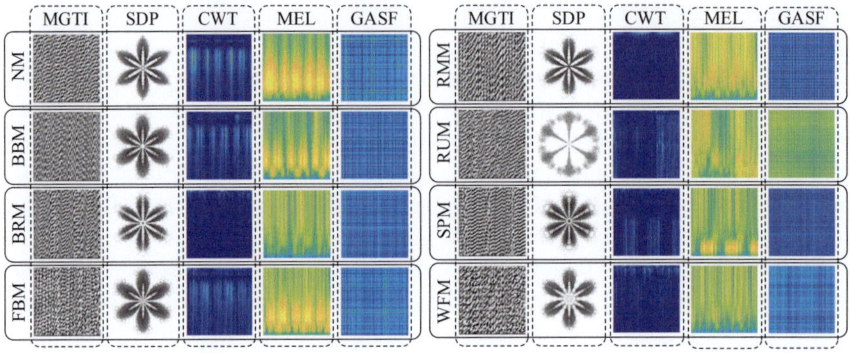

Fig. 7 MGTI compared with other images

multi-scale gray texture image method, 250 sample images are generated from the uniaxial vibration signal, and 2000 sample data are accumulated for 8 fault types. The same number of sample images are generated by other conversion methods. The dataset is divided into training set and test set, and the ratio is 4:1.

In this experiment, Adam is selected as the training optimizer. The initial learning rate is fixed at 1e-3, and the model training undergoes a total of 30 rounds of iterations. The computing platform uses Intel i7-12650 h processor and NVIDIA RTX4060 graphics card (8 GB of video memory), with a hardware environment of 32 GB DDR4 memory, and builds an experimental system based on python 3.8 deep learning framework.

4.2 Ablation Experiment

This ablation experiment systematically verified the contribution of each component to the performance of the model by gradually introducing different modules. In the experimental setup, the basic model B1 uses ResNet combined with MGTI framework, and on this basis, SAM (B2) and CAM (B3) are added respectively to form a comparative experiment. In addition to the above differences, the rest are completely consistent. Table 2 shows the experimental results and the optimal results have been highlighted.

The basic model B1 has shown a high level in core indicators such as Accuracy (97.25%), Recall (97.25%) and F1-score (96.25%), indicating that the framework of ResNet combined with MGTI has good feature extraction ability. After

Table 2 Ablation experimental results (%)

Method	Accuracy	Precision	Recall	F1-score
B1	97.25	97.42	97.25	96.25
B2	98.50	98.58	98.50	97.5
B3	98.75	98.81	98.75	97.75
CBRN + MGTI	**99.75**	**99.52**	**99.75**	**99.50**

the introduction of SAM to form the B2 model, all indicators were significantly improved, including the Accuracy and Recall rate increased by 1.25 percentage points to 98.50%, and the F1-score increased by 1.25 percentage points to 97.5%. This shows that SAM can effectively enhance the attention of the model to the characteristics of key areas and improve the accuracy of classification decision. After the B3 model was replaced by CAM module on the basis of maintaining the advantages of SAM, the Accuracy and Recall rate were further improved to 98.75%, and F1-score reached 97.75%, indicating that the channel attention mechanism can more finely distinguish the importance of different feature channels, so as to obtain better feature representation.

The CBRN + MGTI model proposed in this paper achieves a fault diagnosis Accuracy of 99.75%, significantly surpassing all comparative models. This breakthrough comes from the improvement of CBAM module on ResNet architecture, which makes it have more powerful feature learning ability. The comparison of confusion matrix in ablation experiment is shown in Fig. 8. Combined with the comparison diagram of confusion matrix, the model proposed in this paper can completely eliminate the misclassification among various categories and achieve the theoretical optimal state. Based on the above results, we can conclude that each strategy is necessary.

4.3 Comparative Experiment

Comparative experiment of different image methods. This section verifies the outstanding advantages of the proposed scheme through comparative experiments of different image conversion methods. In addition to converting the uniaxial signal to multi-scale gray texture image (MGTI), it also converts the uniaxial signal to symmetric dot pattern (SDP), continuous wavelet transform (CWT), Mel (MEL) and Gramian angular summation field (GASF). The images obtained by different conversion methods are used as data input. All methods use CBRN model for image classification and fault diagnosis, and ensure that the training parameters and training environment are consistent.

Figure 9 shows the experimental results of this study. At the same time, Figs. 10 and 11 respectively show the visualization results of t-SNE features and the comparison of confusion matrices in the comparative experiment. The analysis results show that the proposed MGTI conversion method performs well in fault diagnosis tasks. From

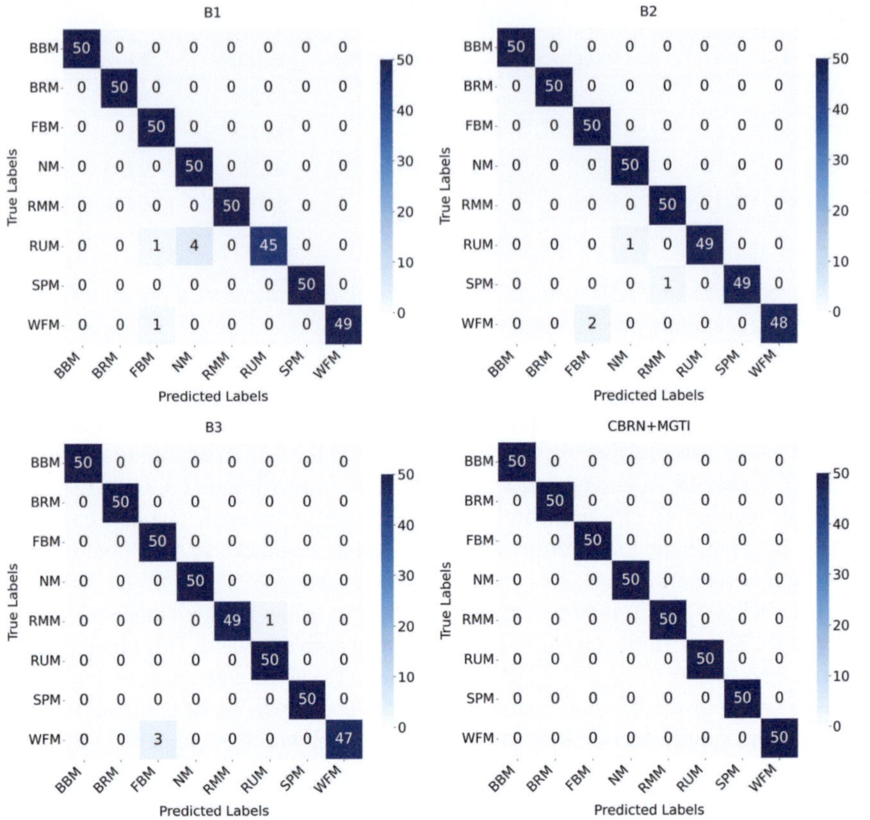

Fig. 8 Comparison of confusion matrix

the perspective of classification Accuracy, MGTI method has achieved the highest accuracy of 99.75% in four types of bearing fault diagnosis, which is 2.75% higher than MEL (97.00%) method, and is significantly better than CWT (96.50%), SDP (89.00%) and GASF (83.00%). This performance improvement is mainly due to the MGTI method, which effectively captures the time–frequency nonlinear coupling characteristics implicit in the vibration signal through the deep mining of multi-scale texture features. In addition, by comparing the performance of different experiments in Precision, Recall rate and F1-score, it is confirmed that CBRN + MGTI model has stronger ability in motor fault diagnosis.

The visualization results of t-SNE feature further verified the feature separability advantage of MGTI. Compared with the serious overlapping phenomenon of sample points in low dimensional space of other methods, the MGTI transformed feature vector shows a clearer inter class boundary in the embedded space, which indicates that it can extract higher-order features with more discriminant. The comparative analysis of confusion matrix shows that MGTI's texture enhancement can improve the reliability of classification decisions by effectively suppressing noise interference.

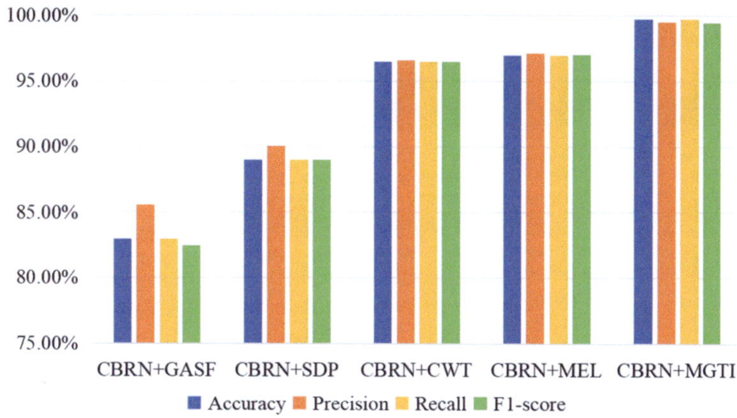

Fig. 9 Comparison of experimental results

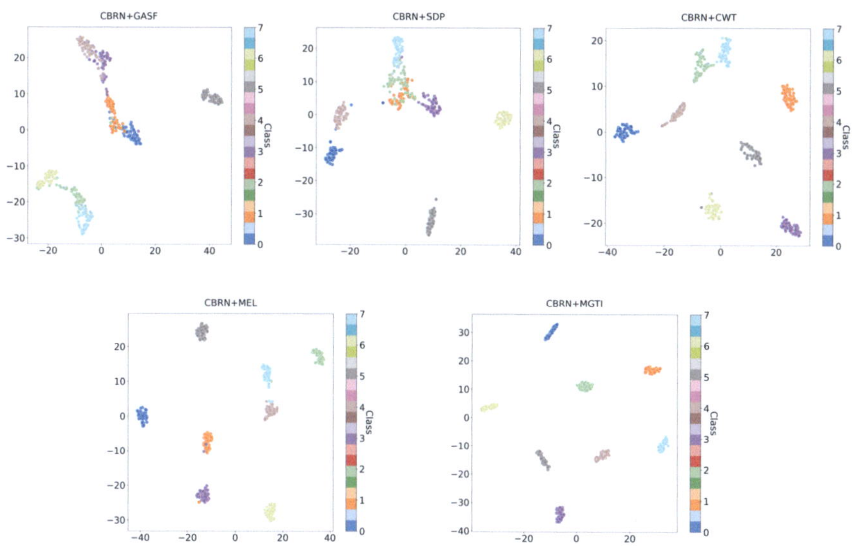

Fig. 10 t-SNE feature visualization of different image conversion methods

Comparative experiment of different diagnostic methods. To further verify the diagnostic advantages of the proposed method, several other methods are selected for comparison. In the ablation experiment, C1 represents the SVM + MGTI method, C2 represents the LSTM + MGTI method, and C3 represents the ResNetSE + MGTI method. Figure 12 shows the experimental results. The ablation experimental results show that the CBRN + MGTI method proposed in this paper shows prominent advantages in accuracy. The accuracy of CBRN + MGTI with 99.75% is much higher than that of other comparison methods, which is 12.25% higher than that of

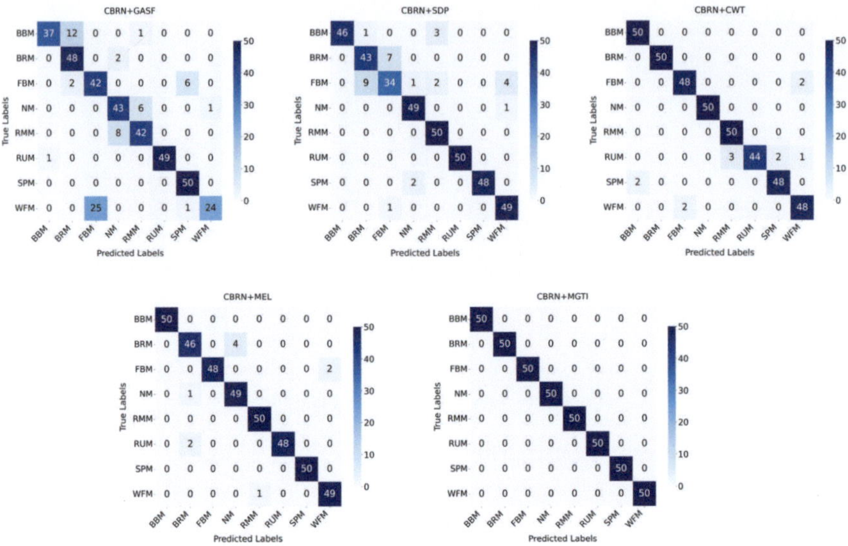

Fig. 11 Confusion matrix results of different image conversion methods

C1, and 5.25% higher than that of C2 and C3, respectively. This fully proves that the traditional machine learning method SVM has limitations in complex feature extraction. Although the LSTM and ResNetSE models based on deep learning can improve the performance, and they are limited by the modeling ability of the network structure for temporal and spatial features. CBRN + MGTI effectively improves the capacity of the model to capture key features and achieves accurate feature fusion and discrimination through innovative multi-scale gray texture images and dual attention mechanism.

Fig. 12 Comparison of experimental results

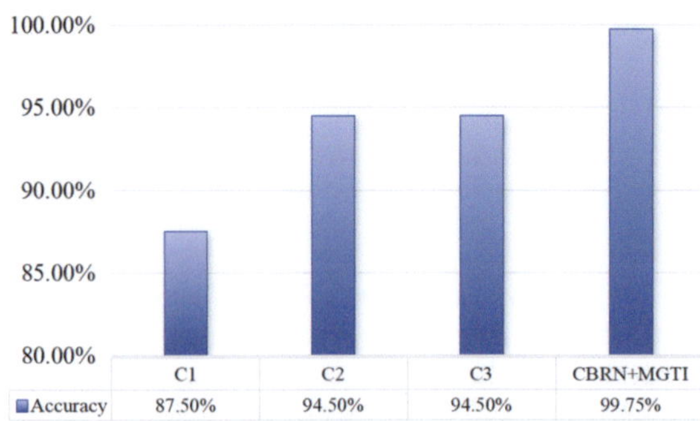

5 Conclusion

Aiming at the new challenges that the vibration signal of the motor surface is disturbed by environmental noise, which makes it difficult to effectively characterize the key features, and the conventional intelligent diagnosis model has insufficient capacity to model multi-scale features and channel-space correlation, a motor fault diagnosis method based on multi-scale gray texture image and dual attention mechanism is proposed.

The proposed method uses vibration signal to construct multi-scale gray texture image (MGTI), and designs an improved residual network (CBRN) based on CBAM module through channel space dual attention mechanism to realize the adaptive focus of key fault features. Through the design of comparative experiments and ablation tests, it is verified that the constructed model can effectively strengthen the texture representation of fault sensitive areas, and its dual attention interaction mechanism significantly improves the identification of motor fault features. Experimental results show that this strategy can maintain excellent accuracy compared with other fault diagnosis methods. The follow-up research will focus on optimizing the model architecture, and is committed to solving the problem of fault diagnosis under the condition of small sample learning and changing working conditions, so as to enhance the generalization ability in practical engineering applications.

References

1. Das AK, Das S, Pradhan AK, Chatterjee B, Dalai S (2023) RPCNNet: a deep learning approach to sense minor stator winding interturn fault severity in induction motor under variable load condition. IEEE Sens J 23(4):3965–3972
2. Feng J et al (2023) A residual multihead self-attention network using multimodal shallow feature fusion for motor fault diagnosis. IEEE Sens J 23(23):29131–29142
3. Wang H, Zheng J, Xiang J (2023) Online bearing fault diagnosis using numerical simulation models and machine learning classifications. Reliab Eng Syst Saf 234:109142. https://doi.org/10.1016/j.ress.2023.109142
4. Zhang T, Liu S, Wei Y et al (2021) A novel feature adaptive extraction method based on deep learning for bearing fault diagnosis. Measurement 185:110030
5. Skowron M, Orlowska-Kowalska T, Kowalski CT (2022) Detection of permanent magnet damage of PMSM drive based on direct analysis of the stator phase currents using convolutional neural network. IEEE Trans Ind Electron 69(12):13665–13675. https://doi.org/10.1109/TIE.2022.3146557
6. Tran MQ, Liu MK et al (2022) Effective fault diagnosis based on wavelet and convolutional attention neural network for induction motors. IEEE Trans Instrum Meas 71:1–13. https://doi.org/10.1109/TIM.2021.3139706
7. Xiao X, Li C, Huang J, Yu T (2024) Rotating machinery fault diagnosis based on one-dimensional convolutional neural network and modified multi-scale graph convolutional network under limited labeled data. Eng Appl Artif Intell 137:109129
8. Ye M, Yan X, Chen N, Jia M (2023) Intelligent fault diagnosis of rolling bearing using variational mode extraction and improved one-dimensional convolutional neural network. Appl Acoust 202:109143

9. Xiang Q, Wang X, Song Y, Lei L (2025) ISONet: reforming 1DCNN for aero-engine system inter-shaft bearing fault diagnosis via input spatial over-parameterization. Expert Syst Appl 277:127248
10. Nie G et al (2025) A novel intelligent bearing fault diagnosis method based on image enhancement and improved convolutional neural network. Measurement 242:116148
11. Zhang X et al (2023) Few-shot fault diagnosis based on heterogeneous information fusion and meta learning. IEEE Sensors J 23(18):21433–21442
12. Zhang X et al (2023) Multisensor-driven cross-domain motor fault diagnosis based on multi-basis energy pattern. IEEE Sens J 23(17):19660–19669. https://doi.org/10.1109/JSEN.2023.3295175
13. Wang Z et al (2025) Few-shot fault diagnosis for machinery using multi-scale perception multi-level feature fusion image quadrant entropy. Adv Eng Inf 63:102972
14. Huang F et al (2023) Demagnetization fault diagnosis of permanent magnet synchronous motors using magnetic leakage signals. IEEE Trans Ind Informat 19(4):6105–6116
15. Liu C, Dong F, Ge K, Tian Y (2024) A new bearing fault diagnosis method based on deep transfer network and supervised joint matching. IEEE Photon J 16(3):1–17, Art no. 8600317. https://doi.org/10.1109/JPHOT.2024.3392392
16. Wen L, Li X, Gao L, Zhang Y (2017) A new convolutional neural network-based data-driven fault diagnosis method. IEEE Trans Ind Electron 65(7):5990–5998
17. Hong-wei F, Ning-ge M, Xu-hui Z, Ce-yi X, Jia-teng M, Yang Y (2024) New intelligent fault diagnosis approach of rolling bearing based on improved vibration gray texture image and vision transformer. Proc Inst Mech Eng C J Mech Eng Sci 238(13):6117–6130. https://doi.org/10.1177/09544062221085871
18. Assad MB, Kiczales R (2020) Deep biomedical image classification using diagonal bilinear interpolation and residual network. Int J Intell Netw 1:148–156
19. Huang Y-J, Liao A-H, Hu D-Y et al (2022) Multi-scale convolutional network with channel attention mechanism for rolling bearing fault diagnosis. Measurement 203:111935
20. Zhang S, Liu Z, Chen Y, Jin Y, Bai G (2023) Selective kernel convolution deep residual network based on channel-spatial attention mechanism and feature fusion for mechanical fault diagnosis. ISA Trans 133:369–383
21. Huang Z, Cao L (2020) Bicubic interpolation and extrapolation iteration method for high resolution digital holographic reconstruction. Opt Lasers Eng 130, Art. no. 106090

Online Condition Monitoring Method for Stator Winding Insulation of Multi-phase Electrical Machines

Dayong Zheng, Geye Lu, Kai Zheng, Yifu Ren, Yuemao Dang, and Pinjia Zhang

Abstract Multi-phase electrical machines are widely adopted in high-capacity propulsion systems, such as those in electric ships, due to their advantages in torque performance and fault tolerance. However, reliable operation requires effective online monitoring of stator winding insulation to anticipate latent short-circuit faults. While partial discharge monitoring is commonly used in grid-connected generators, it is susceptible to interference from inverter switching transients in drive systems. Alternatively, leakage current measurement methods leverage inverter-generated harmonics as excitation to assess insulation impedance without interference. Existing methods, however, are primarily designed for three-phase machines and are not directly applicable to open-end multi-phase configurations. This paper proposes a novel online insulation condition monitoring method tailored for open-end multi-phase machines. The approach uses a common-mode current sensor encircling all phase conductors in opposite directions to cancel load current components and isolate leakage current related to insulation health. Experimental results from a fifteen-phase induction motor demonstrate the method's high accuracy in detecting changes in insulation capacitance and conductance under various operating conditions, confirming its effectiveness for online prognosis and fault prevention.

D. Zheng (✉) · G. Lu
School of Electrical Engineering, Beijing Jiaotong University, Beijing, China
e-mail: 10283@bjtu.edu.cn

G. Lu
e-mail: 10166@bjtu.edu.cn

K. Zheng
Chaoyang Power Supply Company, State Grid Liaoning Electric Power Supply Co., Chaoyang, China

Y. Ren · Y. Dang · P. Zhang
Department of Electrical Engineering, Tsinghua University, Beijing, China
e-mail: dym23@mails.tsinghua.edu.cn

P. Zhang
e-mail: pinjia.zhang@mail.tsinghua.edu.cn

G. Mu et al. (eds.), *Proceedings of the 2nd Electrical Artificial Intelligence Conference, Volume 1*, Lecture Notes in Electrical Engineering 1587,
https://doi.org/10.1007/978-981-95-7932-7_5

Keywords Multi-phase electrical machine · Insulation condition monitoring · Leakage current measurement · Common-mode current · Online prognosis

1 Introduction

The multi-phase electrical machine is widely used in electrical ships for large capacity propulsion application [1, 2]. It has the advantages of low torque pulsation and strong fault tolerance ability, which is essential to meet the demand of high reliability in ship. Although the multi-phase electrical machine can operate under short-circuit fault of one phase winding, it still has the need for online condition monitoring method for insulation condition, which can give prognosis for the latent short-circuit fault. The online condition monitoring method can inform the engineer to maintain the under-test machine timely to avoid the fault-tolerant operation, which will influence the operation performance of the machine.

In the existing online condition monitoring methods, the partial discharge monitoring method is widely used in large generators which connect to the grid directly [3]. However, for inverter-fed machines, the switching transients from the inverter are strong interference for partial discharge monitoring [4]. In contrast, leakage current measurement methods are proposed in recent years [5–8]. This method uses the switching transient harmonics produced by the inverter as the excitation and measure the corresponding leakage current harmonics. The impedance of the insulation can be measured online by this method, which cannot be affected by the switching noise.

However, the most existing studies focus on the online monitoring of three-phase electrical machines. There are few methods for multi-phase winding electrical machines, especially for open-end winding ones. The authors found that the existing leakage current measurement method cannot be applied to the open-end multi-phase electrical machine directly. Therefore, a monitoring method for this kind of machine is proposed and verified by online experimental results.

2 Online Condition Monitoring Method for Winding Insulation

2.1 The Effect of Aging on Insulation

Numerous studies have been conducted on the aging characteristics of stator insulation materials in electric machines. For the pure aging behavior of insulation materials, existing conclusions are relatively well-established. The insulation material can be modeled as a parallel circuit consisting of a resistive component and a capacitive component. Here, the capacitance Ceq represents the capacitive coupling of the insulation, while the resistance Req corresponds to the dielectric losses within the insulation, as illustrated in the figure. During motor operation, the insulation is subjected to multiple stressors including thermal aging, mechanical aging, electrical aging, as well as humidity and contamination. Throughout the aging process, the molecular structure of the insulation undergoes changes—long chains break into shorter ones—leading to alterations in its insulating properties. In the case of thermal aging, the insulation structure tends to develop delamination and pores, which consequently modify the equivalent insulation parameters. The equivalent insulation resistance generally decreases with aging. The change in the equivalent insulation capacitance depends on the specific aging process and mechanism. Typically, the capacitance value decreases during thermal aging. According to accelerated aging test data reported in multiple literature sources, this reduction can reach approximately 30% of the initial value. This established relationship provides a theoretical basis for subsequent online monitoring methods based on terminal electrical signal measurements.

2.2 *Online Monitoring Method for Multi-phase Electrical Machine*

The main problem for online monitoring of insulation condition is that the load current is far larger than the leakage current. For example, the load current can be as large as thousands of amperes but the leakage current is usually several mA. It is very difficult to measure the minor leakage current directly in phase current. The existing current sensors cannot have a accuracy of 10^(−6). Therefore, the influence of the load current must be eliminated in magnetic path and only the flux of leakage current should be measured by current sensor.

In order to eliminate the flux of the load current, the symmetry of the multi-phase system is utilized. The most of the load current is positive sequence components, which means that the sum of all load currents in a group of phase windings is zero. But due to the discrete switching states of inverter, there are common-mode components in the output voltage of the inverter which drives the motor. Under the excitation of common-mode voltage, there exists common-mode leakage currents in stator winding, which can reflect the groundwall insulation health condition. Therefore, the common-mode current can be measured by encircling all phase winding in the same direction using one current sensor as shown in the Fig. 1, taking three-phase electrical machine as an example.

However, for open-end multi-phase electrical machines, the method above cannot be applied because there exists common-mode current from one end to the other of phase winding. For example, the measurement method in the Fig. 2 is wrong because there are common-mode current flowing from inverter 1 to inverter 2, which has no relationship with the insulation health condition. In order to measure the common-mode insulation leakage current, all the inlet and outlet windings in all phase windings should be encircled by one current sensor in different directions, as

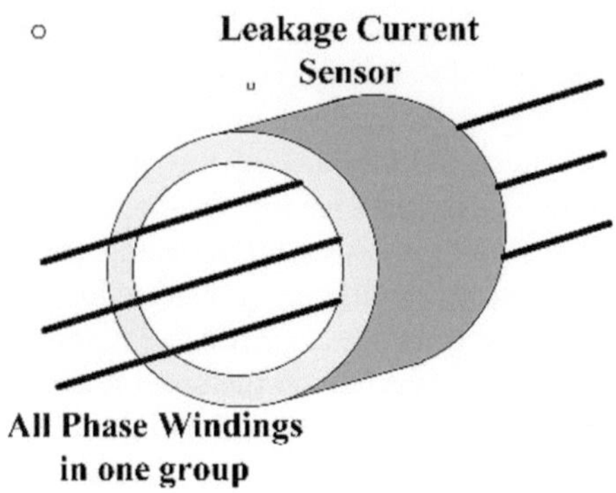

Fig. 1 The leakage current measurement for three-phase electrical machine

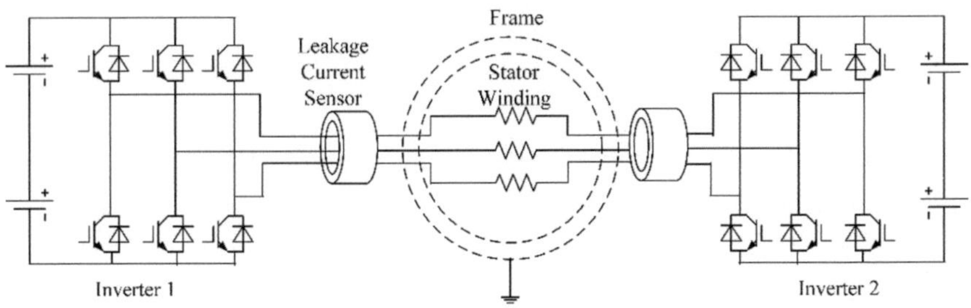

Fig. 2 The false leakage current measurement method for open-end machine

shown in the Fig. 3. In this way, the flux of the common-mode current from inverter 1 to inverter 2 is eliminated, and only the common-mode current flowing through the groundwall insulation is measured, which can reflect the health condition of groundwall insulation.

3 Experiment

3.1 Experimental Setup

The experimental fifteen-phase induction motor features a stator winding composed of three independent five-phase windings in an open-end configuration. Each phase is powered individually by an H-bridge inverter. For one of these five-phase windings, a common-mode measurement method is employed: both the incoming and outgoing cables of the five-phase winding are passed through the center of a current sensor, with the direction of the incoming and outgoing cables opposite to each other. This setup enables online monitoring of the main insulation condition between the stator winding and the ground. During the experiment, the DC bus voltage of the fifteen-phase inverter was set to 150 V (Figs. 4 and 5).

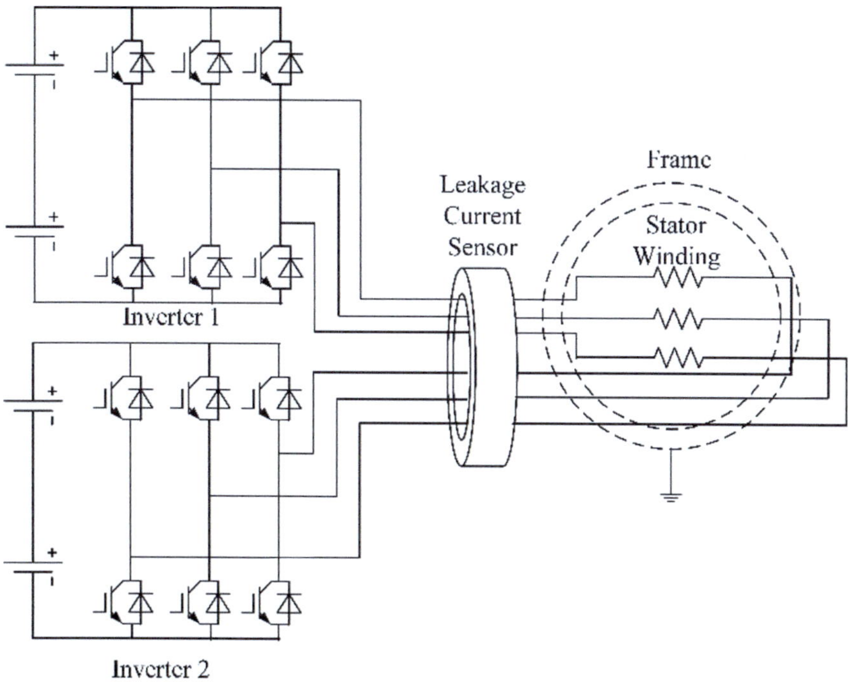

Fig. 3 The correct leakage current measurement method for open-end machine

Fig. 4 The leakage current measurement for 15-phase induction electrical machine

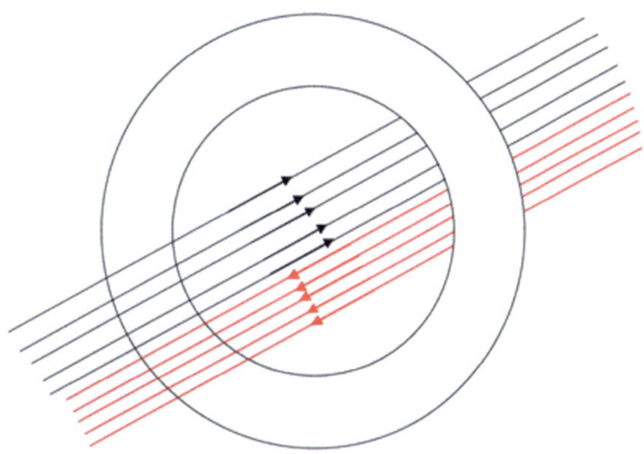

3.2 Experimental Results

The waveforms of the stator winding-to-ground voltage and the measured common-mode leakage current obtained from the test. The results indicate that the ground voltage is a square wave signal with small-amplitude overshoot, while the insulation leakage current manifests as pulse signals corresponding to the edges of the square

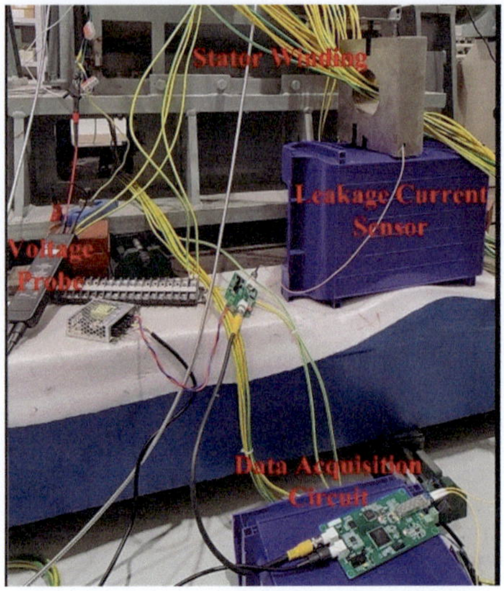

Fig. 5 Experiment platform for leakage current measurement

wave steps. Spectral analysis of the measured signals reveals clear and prominent spectral lines at the switching frequency and its harmonics.

First, under healthy conditions, the equivalent parameters of the stator winding insulation of the fifteen-phase induction motor were measured. The measurement results show that both the magnitude and phase angle of the measured impedance exhibit high resolution. For voltage and current sensors, the most important performance metrics are the ratio error and phase error, which are directly reflected in the measured impedance magnitude and phase angle. The resolution of the magnitude is within 0.1% (corresponding to the ratio error), and the phase angle resolution is within 0.04° (corresponding to the phase error). Since the capacitive leakage current is much larger than the resistive leakage current, the resolution of capacitance measurement is significantly higher than that of resistance (conductance) measurement. Existing studies on insulation material aging have shown that insulation health monitoring and remaining useful life prediction can be achieved based solely on insulation capacitance. Therefore, resistance resolution is not critical for online monitoring purposes (Table 1).

A capacitor with a nominal value of 1.5 nF was connected in parallel to the stator insulation, and online measurements were carried out. The measurement results are presented in the table below. During the monitoring process, steady-state measurements were taken at rotational speeds of 100 rpm (No. 1–5) and 500 rpm (No. 9–11), respectively, and dynamic monitoring was performed during acceleration from 100 to 500 rpm (No. 6–8). The monitoring results demonstrate that the online measurement system maintains high resolution and accurately identifies the incremental change in equivalent capacitance as 1.5 nF (Table 2).

The test results for the fifteen-phase induction motor are summarized in the Table 3. It should be noted that the capacitance and resistance values of the connected

Table 1 Online monitoring results for healthy 15-phase induction electrical machine

No.	Ucm		Icm		Impedance			
	Amp/V	Phase/°	Amp/mA	Phase/°	Amp/Ω	Phase/°	G/nS	C/nF
1	73.02	126.53	2.79	−144.70	26,204.59	−88.77	817	4.05
2	71.65	−42.54	2.73	46.26	26,239.76	−88.81	795	4.04
3	71.59	−32.21	2.73	56.56	26,184.98	−88.77	819	4.05
4	71.60	−169.97	2.73	−81.11	26,205.92	−88.86	758	4.05
5	71.58	19.12	2.73	107.94	26,178.73	−88.83	782	4.05

Table 2 Online monitoring results for aged 15-phase induction electrical machine under different rotation speeds

No.	Ucm		Icm		Impedance			
	Amp/V	Phase/°	Amp/mA	Phase/°	Amp/Ω	Phase/°	G/nS	C/nF
1	82.08	130.97	4.30	−140.00	19,101.31	−89.03	884	5.55
2	82.15	−179.64	4.30	−90.59	19,123.60	−89.05	863	5.55
3	82.08	−2.69	4.29	86.37	19,116.50	−89.06	857	5.55
4	82.12	−79.81	4.30	9.24	19,116.17	−89.05	869	5.55
5	82.16	150.75	4.30	−120.17	19,117.89	−89.08	843	5.55
6	72.91	−178.72	3.81	−89.69	19,135.32	−89.03	883	5.54
7	70.65	55.02	3.69	144.05	19,149.14	−89.03	881	5.54
8	67.41	−40.68	3.52	48.36	19,155.14	−89.04	875	5.54
9	57.95	−111.66	3.02	−22.59	19,180.08	−89.06	851	5.53
10	58.38	−9.77	3.02	79.23	19,305.30	−88.99	909	5.50
11	58.22	−141.45	3.02	−52.41	19,250.20	−89.04	871	5.51

insulation modules are based on component nominal values rather than actual measured values. The designed monitoring system effectively reflects changes in insulation parameters simulated by the parallel insulation modules, mimicking insulation aging: as the nominal capacitance increases, the monitored capacitance value also increases, generally aligning with the nominal value; similarly, as the nominal conductance increases, the monitored conductance value correspondingly rises, largely consistent with the nominal conductance. Therefore, the proposed monitoring method proves effective for application on the fifteen-phase induction motor.

Table 3 The online monitoring results of 15-phase induction electrical machine under different health conditions

No.	Health condition		Impedance			
	ΔC/nF	ΔG/S	Amp/Ω	Phase/°	G/μS	C/nF
1	0	0	26,203	−88.81	0.794	4.05
2	1.5	0	19,159	−89.04	0.872	5.54
3	3	0	14,916	−89.16	0.977	7.11
4	0	1.00E−05	25,152	−73.83	11.1	4.05
5	0	2.00E−05	22,983	−61.03	21.1	4.04

4 Conclusion

This paper proposed and validated a novel online monitoring method for stator winding insulation in open-end multi-phase machines. To address the limitation of conventional methods, which are invalidated by common-mode currents flowing between inverters, a new sensing strategy was introduced. This technique involves encircling all phase conductors in opposite directions with a single current sensor, effectively canceling interference from load currents and isolating the leakage current related to groundwall insulation health. Experimental results from a fifteen-phase induction motor demonstrated the high accuracy of the method, with minimal magnitude and phase errors. The system successfully detected simulated aging through precise tracking of changes in insulation capacitance. The proposed method provides an effective solution for the online prognosis of insulation condition, enabling timely maintenance to prevent faults in critical applications such as marine propulsion systems.

References

1. Song P, Li W, Li Z, Kar NC (2023) Comparative analysis of noise and vibration for dual three-phase IPMSM under healthy and multi-phase open-circuit fault operations. IEEE Trans Magn 59(11):1–7. https://doi.org/10.1109/TMAG.2023.3288831
2. Li R et al (2024) Influence of winding structure on electromagnetic vibration of dual-three-phase permanent magnet synchronous machines. IEEE Trans Ind Appl 60(1):289–300. https://doi.org/10.1109/TIA.2023.3346355
3. Stone GC, Stranges MKW, Dunn DG (2016) Common questions on partial discharge testing: a review of recent developments in IEEE and IEC standards for offline and online testing of motor and generator stator windings. IEEE Ind Appl Mag 22(1):14–19. https://doi.org/10.1109/MIAS.2015.2458337
4. Stone GC (2012) A perspective on online partial discharge monitoring for assessment of the condition of rotating machine stator winding insulation. IEEE Electr Insul Mag 28(5):8–13. https://doi.org/10.1109/MEI.2012.6268437
5. Huang S et al (2024) Diversified assessment of ground-wall insulation in inverter-fed motors by using transient characteristics of leakage current. IEEE Trans Power Electron 39(10):13783–13794. https://doi.org/10.1109/TPEL.2024.3418336

6. Niu F et al (2024) A novel method for monitoring stator interturn insulation degradation of inverter-fed motors based on impedance spectrum division. IEEE Trans Ind Electron 1–13. https://doi.org/10.1109/TIE.2024.3497310
7. Li H, Gu Y, Xiang D, Zhang P, Yue P, Cui Y (2021) Online condition monitoring of line-end coils insulation for inverter-fed machine by switching oscillation mode decomposition. IEEE Trans Ind Electron 69(11):11697–11708. https://doi.org/10.1109/TIE.2021.3120491
8. Xiang DW, Li H, Yan H, Zheng Y, Zhao N, Liu BW (2021) Online monitoring of incipient turn insulation degradation for inverter-fed machine using sensitive tail component in PWM Switching oscillations. IEEE Trans Power Electron 36(8):8730–8742. https://doi.org/10.1109/Tpel.2021.3049809

Research on a Multi-indicator Reconfigurable Equalization Strategy for Lithium-Ion Batteries Based on Hardware-in-the-Loop

Shirui Zhang, Yumiao Jin, Yuhao Pan, Jinhao Meng, Zhengxiang Song, and Kena Chen

Abstract With the growing deployment of lithium-ion battery packs in electric vehicles and energy storage systems, effective equalization control has become essential to ensure safety, consistency, and service life. This study proposes a multi-indicator reconfigurable equalization strategy that integrates voltage and State of Charge (SOC) information to overcome the limitations of conventional voltage-based balancing methods. A system-theoretic control framework is developed, combining segmented decision logic and dynamic current adjustment to achieve adaptive balancing under varying operating conditions. The proposed control algorithm is implemented and validated on a Hardware-in-the-Loop (HIL) platform comprising a battery simulator, equalizer, and controller. Experimental results under diverse imbalance scenarios demonstrate that the hybrid strategy effectively suppresses pseudo-equalization during voltage plateau phases and improves balancing precision by up to 5.5% compared with conventional multi-level voltage deviation control. The results confirm that incorporating energy-state indicators enables more accurate and adaptive balancing control. This work provides practical insights for the design of intelligent Battery Management Systems (BMS) and offers a feasible pathway toward enhancing the efficiency and reliability of next-generation energy storage systems.

Keywords Lithium-ion battery · Reconfigurable equalization control · Battery management system · Hardware-in-the-Loop

S. Zhang · Y. Jin · Y. Pan · J. Meng (✉) · Z. Song
School of Electrical Engineering, Xi'an Jiaotong University, Xi'an, China
e-mail: jinhao@xjtu.edu.cn

K. Chen
State Grid Sichuan Electric Power Research Company Institute, Chengdu, Sichuan, China

Sichuan Key Laboratory of Power Internet of Things, Chengdu, Sichuan, China

G. Mu et al. (eds.), *Proceedings of the 2nd Electrical Artificial Intelligence Conference, Volume 1*, Lecture Notes in Electrical Engineering 1587,
https://doi.org/10.1007/978-981-95-7932-7_6

1 Introduction

With the rapid growth of global energy demand and the increasing urgency of carbon reduction, lithium-ion batteries have become a cornerstone of modern electrification owing to their high energy density, long cycle life, and environmental friendliness [1]. However, the low output voltage of a single cell (2.5–4.2 V) is insufficient for high-voltage applications such as electric vehicles and grid-scale energy storage systems. Therefore, cells are commonly connected in series to form battery packs, which inevitably introduces inconsistencies among cells due to manufacturing tolerances, temperature gradients, and aging effects [2]. These inconsistencies result in unbalanced voltages and State of Charge (SOC), accelerating degradation and posing safety risks during operation [3].

To ensure safe and reliable performance, the Battery Management System (BMS) must accurately monitor individual cell parameters and perform equalization control. Timely balancing prevents the weakest cell from limiting the overall pack capacity and enhances the system lifespan. Conventional equalization methods are generally divided into passive and active approaches [4], where passive balancing dissipates excess energy through resistors, while active balancing redistributes energy among cells using capacitors, inductors, or power converters [5–11].

Despite these advancements, most traditional voltage-based control strategies still face limitations. Voltage alone cannot accurately reflect the internal energy state, especially within the voltage plateau region where the correlation between voltage and SOC becomes weak [8, 12, 13]. Consequently, pseudo-equalization may occur, where the cell voltages appear balanced while their actual stored energies remain inconsistent. This highlights the necessity of integrating energy-state information, such as SOC, into equalization control.

To address this issue, this study proposes a multi-indicator reconfigurable equalization strategy that integrates voltage and SOC information within a unified control framework. The proposed approach employs segmented logic and adaptive current regulation to achieve precise and context-aware balancing under different operating conditions. The strategy is implemented and validated on a Hardware-in-the-Loop (HIL) experimental platform consisting of a battery simulator, equalizer, and controller. Experimental results demonstrate that the proposed method effectively suppresses pseudo-equalization during voltage plateau phases and improves balancing precision compared with conventional multi-level voltage deviation control [9, 14, 15]. This research provides a practical and scalable equalization control framework to enhance the safety, reliability, and energy efficiency of series-connected lithium-ion battery packs.

2 Equalization Strategies

2.1 Equalization Indicators

In this study, a series-connected configuration composed of four identical lithium-ion battery cells is investigated. Based on a detailed analysis of inter-cell inconsistencies, three key parameters—cell capacity, SOC, and terminal voltage—are selected as the primary indicators guiding the proposed equalization strategy. The classification of these indicators is illustrated in Fig. 1.

Terminal voltage and State of Charge (SOC) are adopted as the main control variables. Voltage provides a fast and easily measurable signal, enabling real-time feedback for balancing. However, it is highly sensitive to current, temperature, and polarization, and becomes unreliable in the voltage plateau region, where voltage variations are weakly correlated with the actual stored energy [13, 16].

To compensate for these limitations, SOC is introduced as a complementary indicator that reflects the intrinsic energy imbalance between cells. Unlike voltage, SOC directly represents the available capacity and better captures long-term energy differences. Although SOC cannot be measured directly, it can be effectively estimated in real time through the Coulomb counting method [17, 18], which offers a favorable balance between simplicity and accuracy. The estimation equation is expressed as:

$$SOC(t) = SOC_0 + \frac{1}{C}\int_0^t \eta \cdot I(\tau)d\tau \tag{1}$$

where SOC_0 is the initial SOC, the C_n the rated capacity, η the Coulombic efficiency, and $I(t)$ the instantaneous current

Given the reliable current sensing capability of the HIL system, this method is adopted in the proposed framework for real-time SOC estimation. The combination of voltage and SOC thus enables a more comprehensive and adaptive equalization

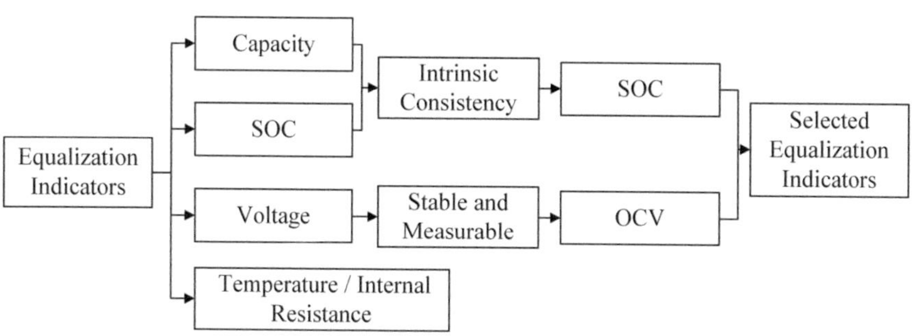

Fig. 1 Classification of equalization indicators

control, effectively addressing pseudo-balancing issues during the voltage plateau phase [19].

2.2 Multi-parameter Hybrid Control Strategy

In light of the complex electrochemical characteristics of lithium-ion batteries and the limitations of single-parameter-based judgment, this study proposes a multi-parameter hybrid control strategy. By constructing a composite evaluation framework that integrates voltage difference and SOC, the strategy aims to overcome the inherent limitations of conventional single-parameter approaches. This enables more accurate identification and dynamic response to imbalances in battery packs.

Voltage, as a surface-level and directly measurable parameter, is significantly affected by internal resistance, temperature, and charge/discharge rates. Particularly in regions of high or low SOC, voltage exhibits pronounced nonlinear behavior. In contrast, SOC—used here as a deep-level capacity indicator—can effectively detect "pseudo-voltage balance" in the voltage plateau region and exhibits heightened sensitivity to capacity disparities within the mid-SOC range (20–80%).

Based on this insight, the proposed method combines layered judgment and dynamic switching to construct a multi-dimensional monitoring framework that covers the entire operating range of the battery system.

The overall structure of the multi-parameter hybrid equalization strategy is shown in Fig. 2. In the mid-SOC region ($20\% < SOC_{avg} < 80\%$), lithium-ion batteries operate within the voltage plateau, where the voltage variation rate is less than 0.05 V per 10% SOC. In this range, the terminal voltage becomes insensitive to changes in stored energy, making SOC difference a more reliable indicator for balancing decisions. Therefore, SOC is adopted as the primary judgment criterion, and equalization is triggered when the SOC deviation exceeds the threshold SOC_{th}. During charging, additional current is supplied to lower-SOC cells to replenish energy, whereas during discharging, higher-SOC cells are discharged more intensively to release excess energy.

When the battery operates in high-SOC or low-SOC regions ($SOC_{avg} < 20\%$ or $SOC_{avg} > 80\%$), it enters a strongly nonlinear phase in which voltage provides clearer indications of imminent overcharge or overdischarge, evidenced by a rapid rise at the end of charging and a pronounced drop before the discharge cutoff. Under these conditions, the strategy switches from SOC-based control to multi-level voltage deviation determination. To ensure operational safety, the cell voltage is constrained within the 3.11–4.18 V range. A three-level voltage deviation structure is retained, where light imbalance corresponds to a threshold of 0.05 V, moderate imbalance to 0.12 V, and severe imbalance to 0.20 V; these thresholds map to balancing currents of 0.8, 1.2, and 1.6 A, respectively, enabling adaptive acceleration of the convergence process under different initial inconsistency levels. Based on the consistency characteristics of the tested battery pack, the SOC imbalance trigger value is set

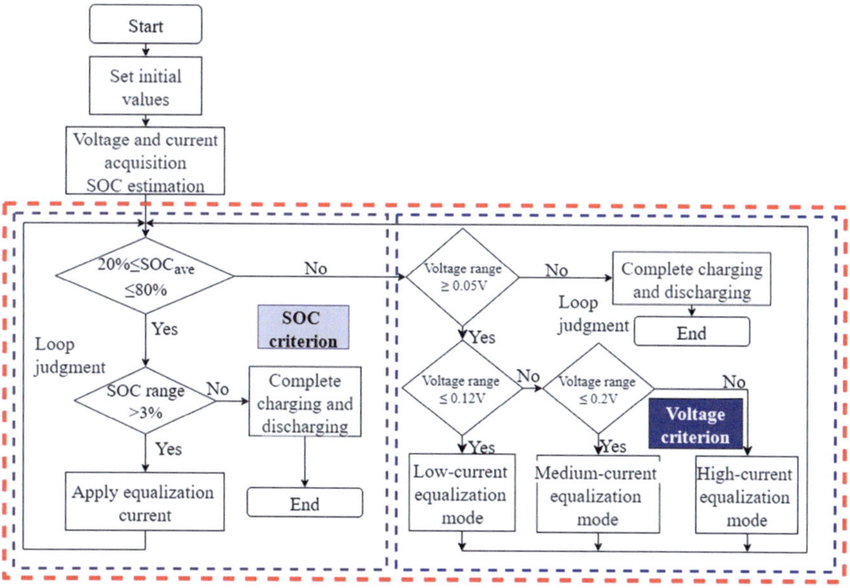

Fig. 2 Flowchart of multi-parameter hybrid control strategy

to $SOC_{th} = 3\%$, ensuring effective balancing in the plateau region and avoiding voltage-dependent misjudgment.

Overall, SOC and terminal voltage are jointly selected as complementary control indicators, with SOC estimated using the ampere-hour integration method. Single- and multi-level voltage deviation strategies provide the baseline structure, while the proposed hybrid strategy is specifically designed to overcome the limitations of voltage-only control in the plateau region and to enhance balancing accuracy across the entire SOC range.

3 Experimental and Analysis

This chapter provides an overview of the experimental platform and setup, followed by a detailed discussion and analysis of the experimental results.

3.1 Experimental Platform

This study adopts the lithium-ion battery model ICR18650/15P, with the parameters listed in Table 1. The battery model is derived using data from Hybrid Pulse Power Characterization (HPPC) tests.

Table 1 Battery parameters

Parameter	Value	Parameter name	Value
Rated capacity	1500 mAh	Charging voltage	4.20 V
Minimum capacity	1450 mAh	Charging current	0.5 C
Nominal voltage	3.6 V	Charging method	CC-CV

HIL simulation, as a testing methodology, is commonly applied in the development of embedded systems or control systems. The core principle involves connecting the controller or hardware under test—representing a portion of the system—to a simulated system environment. Through real-time simulation of the external system's dynamic behavior using dedicated tools, the method enables performance validation of the hardware under realistic operating conditions. In this study, HIL simulation is employed to verify the effectiveness of the proposed balancing strategy.

In this study, the balancing strategy model is developed in Simulink, compiled, and then deployed to VeriStand. The battery simulation controller communicates with the balancer via the CAN bus, enabling direct control of the battery simulator for real-time charge and discharge experiments. During operation, the balancer continuously collects real-time feedback and makes control decisions accordingly.

Figure 3 shows the physical layout of the lower-level cabinet in the complete system, which primarily consists of three components: the battery simulator, the balancer, and the battery simulation controller.

3.2 Analysis of Experimental Results

In this subsection, a comparative analysis of two balancing strategies is conducted based on charge–discharge experiments involving four simulated battery cells. The single-level voltage difference strategy is referred to as Strategy I, while the multi-parameter hybrid control strategy is designated as Strategy II.

During both low SOC and high SOC phases, the balancing behavior in both strategies is governed by multi-level voltage difference control. To enable a meaningful comparison, three distinct sets of initial conditions were configured to cover the voltage plateau region as well as its boundary areas. This experimental design aims to validate the superior performance of the multi-parameter hybrid control strategy.

Low SOC to Voltage Plateau Onset

During the charging experiment, the initial cell voltages were set to 3.41, 3.43, 3.53, and 3.58 V, with an average SOC of 10.676%. The SOC trajectories are shown in Fig. 4, and the corresponding balancing time and SOC deviation are listed in Table 2. Under this condition, Strategy II did not outperform Strategy I in balancing speed and was in some cases slightly slower, as the voltage deviation fell below the threshold early in the plateau region. However, Strategy II achieved much better final balancing accuracy, reducing the SOC deviation by 5.5%.

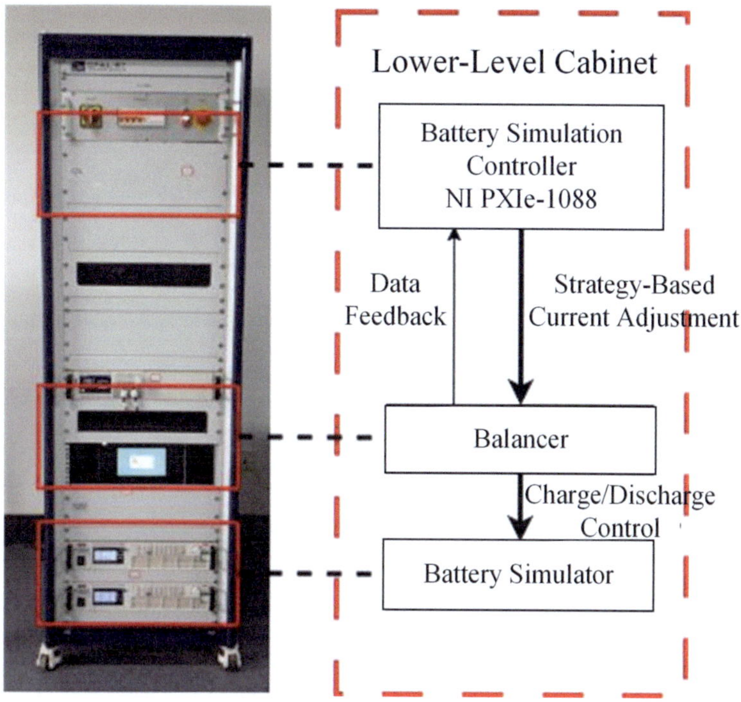

Fig. 3 Hardware cabinet of the embedded system

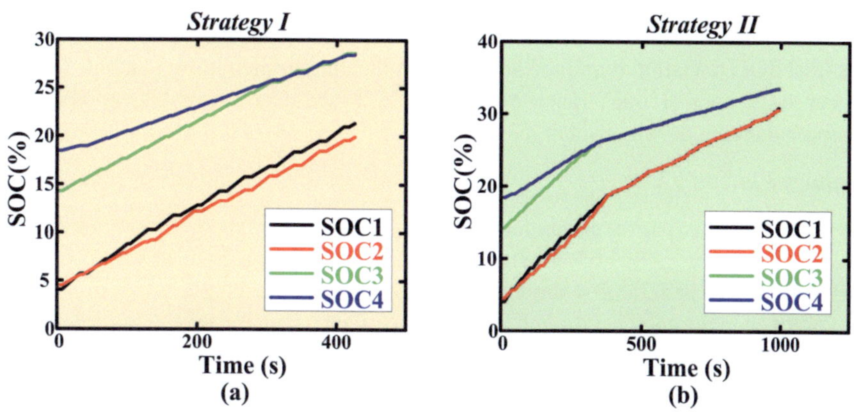

Fig. 4 Charging curves during the low SOC phase

Table 2 Comparison of charging performance

	Strategy I	Strategy II
Balancing time (s)	478	982
Initial SOC deviation (%)	14.3	14.3
Final SOC deviation (%)	8.5	3.0

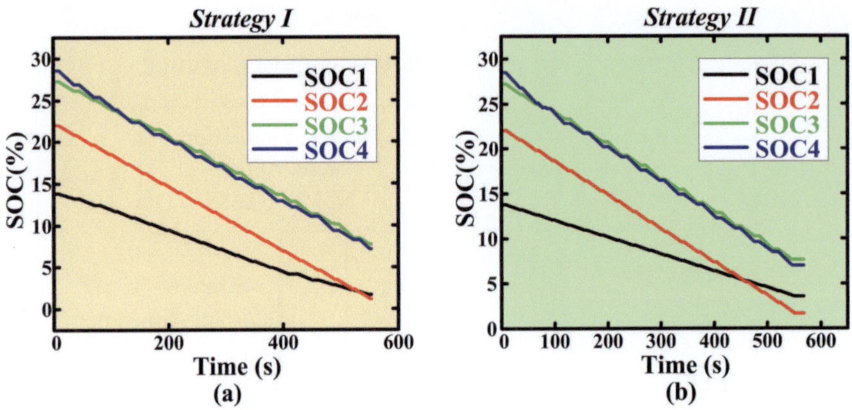

Fig. 5 Discharging curves during the low SOC phase

Table 3 Comparison of discharging performance

	Strategy I	Strategy II
Balancing time (s)	553	540
Initial SOC deviation (%)	14.5	14.5
Final SOC deviation (%)	6.0	4.3

During the discharging experiment, the initial voltages were 3.53, 3.60, 3.64, and 3.66 V, with an average SOC of 23.867%. As shown in Fig. 5 and Table 3, Strategy II shortened the balancing duration by 13 s compared with Strategy I. Nonetheless, its final SOC deviation remained above 3%, indicating limited improvement. This is because the control logic transitions from SOC-based decision-making to voltage-based balancing during discharge.

Voltage Plateau Phase

In the charging experiment conducted within the voltage plateau region, the initial cell voltages were 3.60, 3.65, 3.68, and 3.76 V, with an average SOC of 32.857%. The results in Fig. 6 and Table 4 show that Strategy II achieved a 4.2% lower final SOC deviation than Strategy I, although it required a longer balancing time. This is because voltage changes slowly with SOC in the plateau region, making voltage-based control ineffective for high-precision balancing.

In the discharge experiment, the initial voltages were 4.04, 4.00, 3.96, and 3.90 V, with an average SOC of 72.758%. As shown in Fig. 7 and Table 5. Comparison of Charging Performance, Strategy II again provided higher balancing accuracy, reducing the SOC deviation by 2.5%, but it was slower than Strategy I.

Overall, Strategy II outperformed Strategy I in both charging and discharging during the voltage plateau phase, achieving higher accuracy at the expense of reduced balancing speed (Table 5).

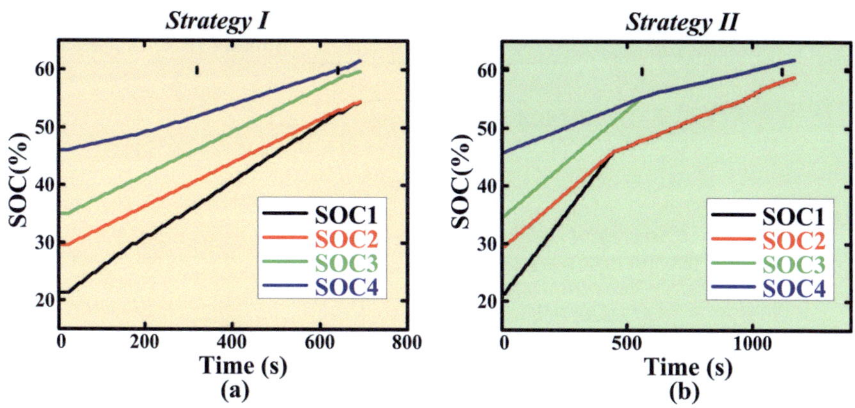

Fig. 6 Charging curves during voltage plateau phase

Table 4 Comparison of charging performance

	Strategy I	Strategy II
Balancing time (s)	666	1150
Initial SOC deviation (%)	24.7	24.7
Final SOC deviation (%)	7.2	3.0

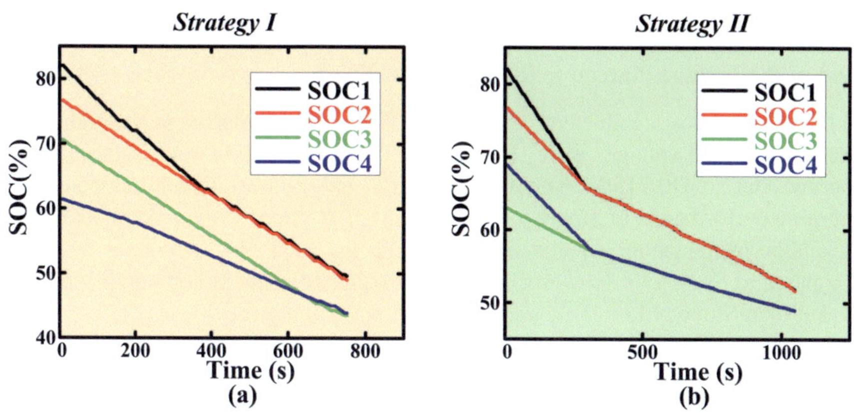

Fig. 7 Discharging curves during voltage plateau phase

Table 5 Comparison of discharging performance

	Strategy I	Strategy II
Balancing time (s)	666	1150
Initial SOC deviation (%)	24.7	24.7
Final SOC deviation (%)	7.2	3.0

Table 6 Comparison of charging performance

	Strategy I	Strategy II
Balancing time (s)	215	227
Initial SOC deviation (%)	11.5	11.5
Final SOC deviation (%)	6.3	3.0

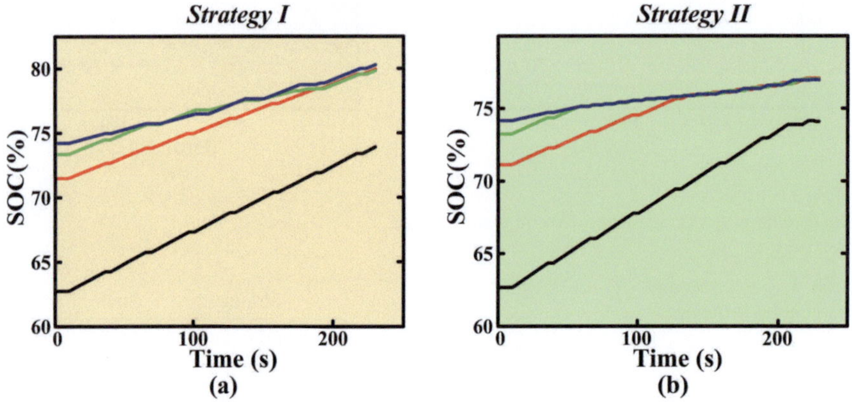

Fig. 8 Charging curves during the high SOC phase

End of the Voltage Plateau to the High SOC Phase

In the charging test conducted from the end of the voltage plateau to the high SOC phase, the initial voltages were set to 3.90, 3.96, 3.98, and 4.00 V, with an initial average SOC of 70.715%. As shown in Fig. 8 and Table 6, which compares the charging performance of Strategy I and Strategy II, Strategy II demonstrated a clear advantage in balancing effectiveness, reducing the final SOC deviation by 3.3% compared to Strategy I. However, in terms of balancing speed, Strategy II was still slower than Strategy I.

In the discharge test conducted from the end of the voltage plateau to the high SOC phase, the initial voltages were set to 4.12, 4.06, 4.00, and 3.96 V, with an initial average SOC of 81.45%. As shown in Fig. 9 and Table 7, Strategy II demonstrated a clear advantage in terms of balancing accuracy and effectiveness, reducing the final SOC deviation by 4.6% compared to Strategy I. However, it still lagged behind in balancing speed. Overall, during the late voltage plateau and high SOC phase, Strategy II achieved better balancing performance.

3.3 *Summary*

Based on the charging and discharging experiments under three representative initial conditions—from low SOC to the voltage plateau and from the plateau to high

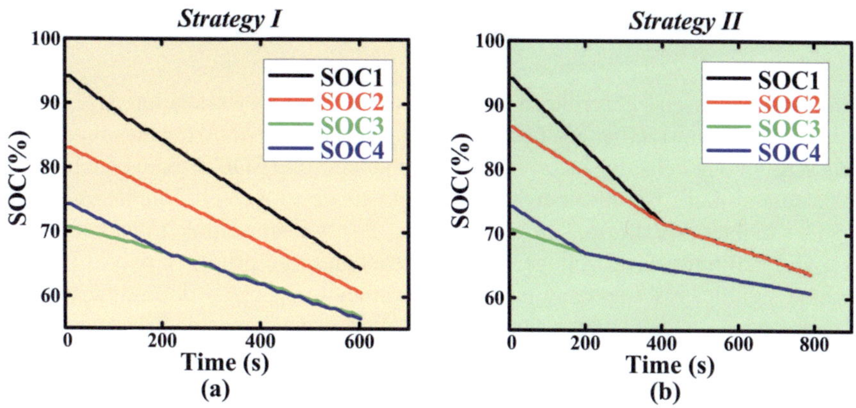

Fig. 9 Discharging curves during the high SOC phase

Table 7 Comparison of discharging performance

	Strategy I	Strategy II
Balancing time (s)	215	227
Initial SOC deviation (%)	11.5	11.5
Final SOC deviation (%)	6.3	3.0

SOC—Strategy II consistently achieved higher balancing accuracy than Strategy I, reducing the final SOC deviation by up to 5.5%. This improvement results from the SOC-based criterion, which remains effective in the plateau region where voltage differences become insensitive. In contrast, Strategy I terminates balancing once the voltage deviation falls below 0.05 V, even when noticeable SOC imbalance remains, leading to residual inconsistency.

Although Strategy II provides higher accuracy, its balancing speed is generally slower due to the flat voltage–SOC relationship in the plateau region. Strategy I therefore maintains an advantage in response speed when voltage deviation is large, whereas Strategy II offers superior final precision by combining SOC and voltage information.

Several practical issues were also identified. Voltage feedback exhibited delays and inaccuracies, affecting control timing. Discharging performance was notably worse than charging, suggesting that current allocation in the hybrid strategy requires further optimization. Additionally, SOC estimation occasionally exceeded physical bounds (below 0% or above 100%), reflecting the inherent drift of the Coulomb counting method and highlighting the need for improved SOC estimation algorithms in future work.

4 Conclusion

This study proposed a generalized equalization control framework tailored for series-connected lithium-ion battery packs, aiming to mitigate inconsistencies in voltage and SOC among cells. By incorporating both voltage and SOC as core equalization indicators, a multi-parameter hybrid strategy was developed and implemented on a HIL simulation platform. The experimental results demonstrated that compared to traditional multi-level voltage deviation strategies, the proposed hybrid control method significantly improved balancing accuracy—especially during the voltage plateau phase—where voltage alone fails to reflect actual capacity disparities.

Specifically, Strategy II reduced SOC deviation by up to 5.5% while ensuring cell safety under various charging and discharging conditions. Despite a trade-off in balancing speed, the enhanced precision of energy distribution validates the necessity of integrating energy-state information in equalization decisions. The hybrid approach effectively avoided pseudo-equalization and minimized residual imbalance.

Nonetheless, some limitations were observed. In particular, the coulomb counting method used for SOC estimation exhibited cumulative errors under prolonged discharge conditions, occasionally resulting in invalid SOC readings (<0% or >100%). Moreover, the control response was sometimes delayed due to feedback inaccuracies from the balancing hardware. These findings suggest that further improvements in current sensing precision, adaptive current allocation, and robust SOC estimation algorithms—possibly incorporating machine learning—are essential in future research.

In conclusion, this research provides a practical, intelligent balancing strategy that enhances battery consistency, operational safety, and energy efficiency. It offers promising insights for the design of next-generation Battery Management Systems (BMS) in electric vehicle and energy storage applications.

Acknowledgements This work was supported by the Science and Technology Project of State Grid Sichuan Electric Power Corporation under Grant 521997230036.

References

1. Lu L, Han X, Li J, Hua J, Ouyang M (2013) A review on the key issues for lithium-ion battery management in electric vehicles. J Power Sources 226:272–288
2. Cheng W (2021) Research on series equalization control strategy of power battery packs (in Chinese), MS thesis, Jiangnan University, Wuxi, China
3. Han W, Zou C, Zhou C, Chen M (2018) Estimation of cell SOC evolution and system performance in module-based battery charge equalization systems. IEEE Trans Smart Grid 10(5):4717–4728
4. Wu Z, Wu S, Zhang C et al (2014) Lithium battery pack equalization circuit and its equalization strategy research (in Chinese). Power Electron Appl 33(11):50–54

5. Li Y, Xu J, Mei X, Zhang C, Song S (2019) A unitized multiwinding transformer-based equalization method for series-connected battery strings. IEEE Trans Power Electron 34(12):11981–11989
6. Shang Y, Zhang Q, Cui N et al (2017) A cell-to-cell equalizer based on three-resonant-state switched-capacitor converters for series-connected battery strings. Energies 10(2):206
7. Manjunath K, Kalpana R, Singh B (2023) A two-stage module-based cell-to-cell active balancing circuit for series-connected lithium-ion battery packs. IEEE Trans Energy Convers 38(4):2282–2297
8. Gao P, Shu Y, Yang J et al (2023) Research on lithium battery equalization strategy based on load state consistency (in Chinese). Comput Simul 40(4):108–113
9. Sun Q, Lv H, Wang S, Gao J, Zhao Z (2021) Optimized state of charge estimation of lithium-ion battery in SMES/battery hybrid energy storage system for electric vehicles. IEEE Trans Appl Supercond 31(8):1–6
10. Liu J (2019) Research on active balancing strategy and SOC estimation for electric vehicle battery packs (in Chinese)
11. Li L, Xu Y, Jiang C (2023) Research on equalization strategy for series-parallel lithium battery pack (in Chinese). Power Technol 47(01):51–56
12. Manjunath K, Kalpana R, Singh B (2023) A two-stage module based cell-to-cell active balancing circuit for series connected lithium-ion battery packs. IEEE Trans Energy Convers 38(4):2282–2297
13. Tian C, Wang W, Zhang F et al Research progress and Prospect of SOC estimation methods for power batteries in electric vehicles (in Chinese). China J Highw Transp 1–26
14. Gao P, Shu Y, Yang J et al (2023) Research on lithium battery equalization strategy based on load state consistency (in Chinese). Comput Simul 40(04):108–113, 138
15. Ma X, Zhang X, Guo Z et al (2025) Design of state estimation method for power battery in new energy vehicles (in Chinese). Intern Combust Engine Parts 07:25–27
16. Wang X, Zhang W, Tian H et al Estimation methods of static and dynamic state of charge for lithium-ion power battery in new energy vehicles (in Chinese). Power Electron Technol 1–6
17. Bao H, Yu Y (2013) SOC estimation error correction of battery based on ampere-hour integral method (in Chinese). Comput Simul 30(11):148–151+159
18. Luo Y, Qi P, Huang H et al (2020) Research on SOC estimation method based on capacity correction and ampere-hour integration (in Chinese). Automot Eng 42(05):681–687
19. Peng Z, Wang S, Xiao S SOC & SOH estimation method for electric vehicle batteries based on SDAE-DCPInformer (in Chinese). J Intell Syst 1–16

Research on the Development of Electric Aircraft for Intelligent Aviation

Yang Li, Zihua Qiu, and Meilun Zhang

Abstract As digitalization transforms industries globally, electric aircraft have emerged as pivotal players in the quest for green aviation, seamlessly integrating with artificial intelligence (AI) to birth the era of intelligent aviation. This paper embarks on a comprehensive exploration of the current state and trajectory of electric aircraft development worldwide. It delves into the primary trends shaping this evolution, alongside the intricate challenges that persist. Our examination highlights critical technological clusters essential for the widespread adoption of electric aircraft. The analysis further underscores how intelligent technologies can bolster these aircraft's performance, reliability, and operational efficiency. However, the transition to intelligent aviation presents a unique set of hurdles. Among these are constraints in onboard computing power and algorithmic proficiency, stringent demands for safety guarantees and airworthiness certifications, as well as complex airspace management and air traffic coordination issues. In addressing these challenges, the paper proposes targeted strategies. It emphasizes the importance of enhancing computational capabilities and refining algorithms for optimized aircraft performance. Moreover, it advocates for robust frameworks ensuring the safety and airworthiness of electric aircraft integrated with AI systems. Lastly, it calls for coordinated efforts to streamline airspace management and air traffic control in an increasingly intelligent aviation landscape. Ultimately, the findings of this paper aim to serve as a guide for researchers, regulatory bodies, and industry stakeholders. By shedding light on critical research areas and policy recommendations, it seeks to expedite the secure and scalable integration of electric aircraft into an AI-enhanced airspace, paving the way for a smarter, greener aviation future.

Keywords Electric aircraft · Intelligent aviation · Smart flight control · Energy management · eVTOL

Y. Li (✉) · Z. Qiu · M. Zhang
AVIC the First Aircraft Institute, Xi'an, China
e-mail: 284037710@qq.com

G. Mu et al. (eds.), *Proceedings of the 2nd Electrical Artificial Intelligence Conference, Volume 1*, Lecture Notes in Electrical Engineering 1587,
https://doi.org/10.1007/978-981-95-7932-7_7

1 Introduction

In the digital era, electric aircraft have assumed a pivotal role in green aviation, marking a significant stride towards intelligent aviation through deep integration with artificial intelligence (AI). This paper undertakes a comprehensive review of the domestic and international landscape of electric aircraft development, examining key trends and the challenges inherent in their evolution. Our exploration delves into the essential technological clusters that must be addressed for wider deployment of electric aircraft. We then assess how intelligent technologies can elevate their performance, reliability, and operational efficiency. However, the emergence of intelligent aviation brings forth its own set of obstacles. These challenges include constraints in onboard computing resources and algorithmic proficiency, stringent requirements for safety assurance and airworthiness certification, and complex issues in airspace management and air traffic coordination. To navigate these hurdles, the paper proposes targeted recommendations. Firstly, it emphasizes the importance of enhancing computational capabilities and refining algorithms to optimize electric aircraft performance. Secondly, it underscores the need for comprehensive frameworks that ensure the safety and airworthiness of AI-integrated electric aircraft. Lastly, it advocates for coordinated efforts to streamline airspace management and air traffic control, enabling smoother operations within the intelligent aviation paradigm. The findings of this paper aim to provide valuable insights to researchers, regulatory bodies, and industry stakeholders. By highlighting priority research directions and policy measures, we seek to accelerate the secure and scalable adoption of electric aviation in an AI-enabled airspace. Our hope is to pave the way for a future where intelligent aviation not only reduces our environmental footprint but also redefines the aviation industry for greater efficiency and safety.

As the aviation industry transitions towards greener operations, sustainable flight solutions have gained widespread consensus. Electric aircraft, with zero emissions, low noise, and high energy efficiency, are transitioning from concept to application, emerging as key drivers of the green aviation revolution [1, 2].

Electric aircraft rely on electricity and electronic systems, integrating flight and propulsion with high energy demands and storage challenges. Advanced safety measures, including software algorithms, ensure redundancy. These characteristics make them well-suited for, and in desperate need of, intelligent technologies for empowerment and enhancement.

This paper focuses on the technological development of electric aircraft in smart aviation. It analyzes key technologies and intelligent applications, identifying challenges and potential solutions for the intelligent transformation of electric aircraft.

2 Electric Aircraft Development

Electric aircraft development is advancing rapidly with diverse technological pathways and specific application scenarios. Notably, significant progress has been achieved in electric vertical takeoff and landing (eVTOL) aircraft and pure electric fixed-wing aircraft. eVTOLs aim to revolutionize urban short-range transportation, addressing traffic congestion with their vertical takeoff and landing, low noise, and zero emissions. Applications include aerial tourism, urban taxi services, emergency rescue, and high-value cargo transport. On the other hand, pure electric fixed-wing aircraft technology is more mature, with some models entering regulatory compliance and mass production stages.

The electric aircraft industry is transitioning from technology development to airworthiness certification and commercialization. With advancements in energy storage, flight control systems, and supportive policies, electric aircraft are poised for large-scale deployment in niche markets, transforming human transportation (Fig. 1).

2.1 *Overview of International Electric Aircraft Development*

Electric technology is seen as a pivotal aspect of aviation's "third technological revolution." In response, Europe and the US have developed multiple electric aircraft plans and invested significantly in R&D, steering the aviation industry towards green, sustainable practices (Fig. 2).

The United States is the most dominant force in global electric aircraft research and development. FAA and DOE are actively promoting research of electric aircraft. As early as 2010, Boeing and NASA began collaborating on electric aircraft technology research, proposing the SUGAR (Subsonic Ultra Green Aircraft Research) project

Fig. 1 Major electric aircraft by year of first flight (2018–2025)

Fig. 2 Joby S4 tiltrotor electric aircraft

Fig. 3 Volocity multi-rotor electric aircraft

to explore hybrid electric propulsion system solutions. Several U.S. startups, such as Joby Aviation and Zunum Aero, are also developing eVTOL and hybrid-powered regional aircraft [3] (Fig. 3).

The European Union holds a leading position in electric aircraft research, continuously funding research on electric aircraft technologies through the "Clean Sky" programs, particularly in areas such as hybrid electric propulsion, aviation electrical systems, and lightweight materials.

2.2 *Current Status of Domestic Electric Aircraft Development*

Despite a later start than the US, domestic electric aircraft research has made significant strides. In 2012, the Ruixiang series from Liaoning General Aviation Academy achieved its maiden flight, marking a technological breakthrough. By 2024, EH216-S secured China's first manned drone license, while the VT-30 electric aircraft enhanced its endurance and automation levels [4] (Fig. 4).

Fig. 4 VT-30 lift-and-cruise eVTOL

2.3 *Trends and Challenges in Electric Aircraft Development*

Electric aircraft are experiencing rapid global development despite challenges such as limited endurance, technological immaturity, and incomplete airworthiness standards. The market recognizes their potential, evidenced by a more than doubling of median investments post-2020. Domestic and European efforts are accelerating to match US progress. Further, observations from the Russia-Ukraine conflict highlight the need for stealth, low noise, cost-effectiveness, and site flexibility in future urban combat drones, making electric aircraft, particularly eVTOLs, well-suited to meet these demands (Fig. 5).

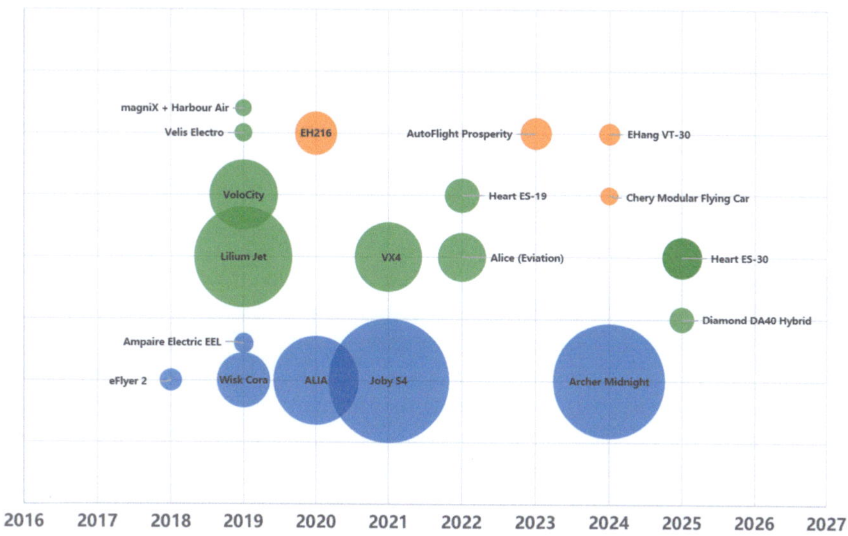

Fig. 5 Investment in electric aircraft by first flight year (2018–2025)

3 Key Technologies for Electric Aircraft

Electric aircraft are an emerging field requiring advancements in key technologies, integrated power-aircraft design, high power-density electric propulsion, energy storage and management, and low-cost lightweight materials.

3.1 Integrated Power-Aircraft Design Technology

Electric aircraft power systems involve more than just replacing traditional engines with electric devices, they require a comprehensive redesign integrating propulsion and aircraft. Without the constraints of traditional fuel engines and tanks, electric aircraft offer greater design freedom. This allows for distributed layouts with high-power motors and electric propulsion systems strategically positioned across the wings, fuselage, or tail. This approach optimizes aerodynamics, propulsion, structures, and systems, significantly enhancing overall aircraft performance.

Enhanced integration of power and aircraft design leads to decreased drag and noise, enhancing flight efficiency, passenger comfort, and overall performance. By optimizing electric propulsion technology, aviation can achieve significant advancements in safety, efficiency, and operational costs [5].

3.2 High Power-Density Electric Propulsion Technology

The importance of the electric propulsion system in electric aircraft cannot be overstated, as it directly influences the performance of the aircraft [6, 7]. A high power-density electric propulsion system is crucial for achieving optimal speed, climb rate, payload capacity, and range. These systems are not only comprised of efficient and lightweight electric motors but also incorporate advanced power electronics, thermal management, and structural integration technologies. The overall performance of these systems greatly impacts the aircraft's energy efficiency, endurance, and economics. In phases of high energy consumption such as vertical takeoff and landing, the advantages of energy density play a crucial role in reducing takeoff and landing times and overall energy consumption. Therefore, high power-density electric propulsion technology is not only a game-changer for improving current electric vehicles but also a key factor in achieving scalability, intelligence, and commercial success. The development of high power-density electric propulsion technology sets the stage for the competitive landscape and future trajectory of the electric aircraft industry. Its advancements are essential for pushing the boundaries of electric flight and realizing the full potential of electric aircraft (Table 1).

Table 1 Stages and targets of motor development

Time period	Typical applications	Energy density target (kW/kg)
2010–2015	NASA/Embraer initiates X-57 project, developing permanent magnet synchronous motors	3–5
2016–2020	Harbour Air eBeaver equipped with magniX magni250 motor achieves first flight	5–10
2021–2025	Joby S4 utilizes 12 kW/kg motors to achieve over 100 h of flight	10–15
2026–2030	High-temperature superconducting (HTS) motors reach 20 kW/kg power density IATA Net Zero roadmap: MEA supports 50-seat medium electric aircraft	15–20
2031–2040	Enables transcontinental flights	>20–30

3.3 *Energy Storage and Comprehensive Energy Management Technology*

Electric aircraft face significant challenges in achieving energy storage efficiency compared to traditional aviation fuels [8, 9]. Current mainstream solutions, such as lithium-ion batteries, have energy densities that are only a fraction of that of aviation Jet-A fuel [10]. However, there is ongoing exploration of emerging energy storage technologies, such as solid-state batteries and hydrogen fuel cells, which have shown promising developments in laboratory prototypes [11, 12]. Solid-state batteries and hydrogen fuel cells, in particular, have demonstrated energy densities that surpass those of traditional lithium-ion batteries. It is projected that by 2035, advanced solid-state batteries could achieve energy densities exceeding 1,000 Wh/kg [13]. These advancements are crucial for addressing the weight and volume constraints that are vital for the engineering realization of electric aircraft. In addition to energy storage, comprehensive energy management technology plays a critical role in optimizing aircraft performance. Real-time monitoring of battery packs, motors, and power electronics, as well as effective energy allocation and thermal management, can help reduce overall energy demands and minimize over-design in traditional aircraft energy systems. Overall, advancements in energy storage and comprehensive energy management technologies will not only define the performance boundaries of electric aircraft but also drive the transition from experimental validation to scaled applications. By addressing these challenges, electric aircraft will be one step closer to becoming a viable and sustainable alternative to traditional aviation fuels (Table 2).

Table 2 Stages and targets of battery development

Time period	Typical applications	Energy density target (kW/kg)
2010–2015	NASA initiates electric aircraft testing using lithium-ion batteries GE completes ground validation of 200 Wh/kg energy storage system	150–250
2016–2020	eVTOL prototype Opener BlackFly uses NCA batteries. EH216 achieves first flight and long-duration hover testing using lithium-ion batteries	200–300
2021–2025	Quantum Scape solid-state batteries reach 350–400 Wh/kg energy density and undergo integration testing with Joby S4 CATL laboratory prototype solid-state batteries achieve 400–500 Wh/kg energy density	300–500
2026–2030	Beta Alia implements practical hydrogen-electric hybrid systems IATA Net Zero roadmap: MEA supports 50-seat medium electric aircraft	500–800
2031–2040	Electric regional aircraft apply metallic lithium solid-state batteries	800–1,200

3.4 Low-Cost Lightweight Materials Technology

The energy density of electric aircraft energy storage systems is not as efficient as traditional aviation fuels, leading to limitations in range and payload capacity. To address this challenge, lightweight materials play a crucial role in reducing structural weight, easing the burden on propulsion systems, and enhancing overall energy efficiency and flight performance. By incorporating high-strength, low-density composite materials and lightweight alloys, aircraft weight can be significantly reduced while still maintaining structural integrity and safety. Studies have shown that carbon fiber composites, for example, can reduce weight by up to 30% compared to aerospace aluminum alloys, resulting in decreased propulsion power requirements and extended flight endurance [14, 15]. In addition to lightweight materials, cost-effective structural design is essential for the widespread adoption of electric aircraft. Optimizing composite material formulations and utilizing easily processed solutions like aluminum alloys or thermoplastic composites can greatly reduce production costs while ensuring structural performance. Furthermore, the use of multifunctional structural materials that combine load-bearing, thermal insulation, and vibration damping functions in airframe components can simplify assembly processes, reduce weight, and cut down on costs. These innovations are key to promoting the use of electric aircraft in short-haul transportation, intercity commuting, and logistics applications where efficiency and cost-effectiveness are paramount.

4 Applications of Intelligent Technologies in Electric Aircraft

The integration of intelligent technologies in electric aircraft is revolutionizing the aviation sector, leading to increased autonomy and efficiency. By 2030, it is expected that 85% of electric aircraft will be equipped with intelligent systems, a significant jump from the current 40% in traditional aircraft [16]. This advancement not only boosts the sustainability and economic feasibility of electric aircraft but also offers various unique benefits in autonomous control, safety measures, and operational reliability. The incorporation of intelligent technologies is reshaping the future of aviation, paving the way for safer, more efficient, and sustainable air travel.

4.1 *Intelligent Flight Control*

The advancement of electric aircraft heavily relies on the utilization of intelligent flight control technology. Unlike conventional aircraft, electric aircraft employ distributed electric propulsion layouts, posing challenges such as aerodynamic coupling and redundancy control that surpass the capabilities of traditional control laws [17, 18]. Therefore, intelligent flight control has transitioned from being an optional feature to a necessary component for ensuring flight safety, enhancing performance, and driving commercial success. A key aspect of safety is the implementation of intelligent fault-tolerant control, which allows for the adaptation of control strategies in the event of partial unit failures. This capability not only ensures safe degradation but also serves as the basis for obtaining airworthiness certification. From an economic standpoint, algorithms like model predictive control (MPC) play a crucial role in optimizing power allocation and flight trajectories in real time. By maximizing energy efficiency, MPC directly contributes to extending the range of electric aircraft. Additionally, intelligent flight control technology supports autonomy in urban airspace operations by enabling autonomous takeoff and landing, obstacle avoidance, and collaborative swarm flight. These advancements are essential for the future development of electric aircraft and pave the way for a more efficient and autonomous aviation industry.

4.2 *Intelligent Energy Management*

Harnessing intelligent energy management technology is crucial for maximizing the efficiency and performance of electric aircraft. By focusing on optimizing the batteries and propulsion systems through sophisticated AI algorithms, these cutting-edge solutions enable dynamic power allocation and predictive charging optimization. This innovative approach is a game-changer in addressing the endurance limitations often faced by electric aircraft. The integration of intelligent energy management technology into electric aircraft architecture has a transformative impact on flight operations. Real-time optimization of power allocation during flight, along with planning the most cost-effective flight profiles, significantly increases the aircraft's range. An exemplary demonstration of this is the AI system implemented in the Beta ALIA cargo eVTOL, which successfully extends its endurance from 250 km to an impressive 320 km by dynamically allocating battery packs based on route wind conditions [19]. Moreover, the incorporation of AI-based battery health prediction management ensures real-time monitoring of cell states, enhancing safety measures by preventing over-discharge and thermal runaway. This proactive approach safeguards flight safety and reliability. Additionally, intelligent scheduling of charging and maintenance requirements optimizes aircraft availability while reducing lifecycle costs. Overall, intelligent energy management technology presents a promising solution for enhancing the efficiency, safety, and operational effectiveness of electric aircraft.

4.3 Intelligent Fault Diagnosis

Intelligent fault diagnosis technology plays a crucial role in safeguarding the safety and reliability of electric aircraft, acting as an essential "immune system" against potential malfunctions. With the integration of multiple high-powered electric drive systems, power batteries, and sophisticated computers, the complexity and interconnections within electric aircraft surpass those of traditional planes. Consequently, new and rapidly spreading fault modes emerge, challenging conventional diagnostic methods based on thresholds and regular checks. Intelligent fault diagnosis relies on cutting-edge deep learning algorithms to process extensive sensor data in real-time, identifying potential failures in motors, electronic controls, or battery packs well in advance. This proactive approach represents a paradigm shift from reactive problem-solving to preemptive detection. For instance, the Pipistrel Velis Electro model can predict motor bearing wear up to 72 h before the issue arises, showcasing the technology's efficacy [20]. Additionally, electric aircraft support over-the-air (OTA) diagnostics, eliminating the need for physical disassembly for troubleshooting. In essence, intelligent fault diagnosis technology is indispensable for ensuring the seamless operation and safety of electric aircraft in today's rapidly evolving aviation landscape [21].

5 Challenges and Solutions for Intelligent Aviation

The integration of intelligent technologies with electric aircraft is revolutionizing the aviation industry, leading to unprecedented gains in efficiency and autonomy. However, this advancement also presents numerous challenges. Issues arise from limited computational resources, rigorous safety regulations in aviation, and the intricate coordination required in low-altitude environments. Overcoming these obstacles is crucial for the successful implementation of these cutting-edge technologies in aircraft systems.

5.1 Computational Resources and Algorithm Efficiency

Cutting-edge technologies in electric aircraft are increasingly reliant on real-time artificial intelligence (AI) processing for tasks like predicting flight paths and diagnosing faults. However, the limited computing resources at the edge often result in decreased algorithm efficiency. For instance, the processors found in typical electric vertical takeoff and landing (eVTOL) aircraft, such as the NVIDIA Jetson series, offer significantly lower performance compared to cloud data centers. This deficiency leads to processing delays exceeding 100 ms during high workloads, impacting the stability of autonomous flights [22]. Furthermore, this bottleneck contributes to increased battery energy consumption, with AI calculations consuming around 15–20% of total power and adding to the thermal management challenges, potentially reducing flight endurance by 10% [23]. In urban air mobility (UAM) settings, these inefficiencies can lead to path optimization failures, which have been observed to rise by 5–8% due to delays in replanning caused by latency issues during wind disturbances. A survey conducted on global eVTOL projects in 2025 revealed that approximately 30% of them encountered compute overload problems, emphasizing the limitations on overall system performance. To tackle these challenges, techniques like model compression can be utilized to enhance AI speed, while hybrid edge-cloud architectures utilizing 5G networks for offloading non-critical tasks have proven successful in enhancing efficiency by 25% [24], as evidenced by the EHang VT-30 project. Additionally, the integration of quantum-assisted edge chips is expected to reduce power consumption and provide more robust computational support for dynamically changing scenarios.

5.2 Safety Assurance and Airworthiness Certification

The use of intelligent technologies in aviation, while promising in terms of fault prediction accuracy, presents a significant challenge when it comes to meeting airworthiness certification standards due to the opaque nature of neural network decisions.

The lack of explainability in these systems often results in certification cycles being prolonged by up to 18–24 months, putting passenger safety and system robustness at risk in critical situations like bird strikes or electromagnetic interference [25]. This issue has had far-reaching implications, as evidenced by the 2025 Lilium Jet certification delay, which has stalled financing at $1 billion [26]. Furthermore, global simulations on electric VTOL accidents have shown that AI black-box issues contribute to 25% of potential risks, leading to a crisis of trust within the industry. To address this challenge, a shift towards explainable AI frameworks, utilizing tools like SHAP values to visualize decision-making processes, is essential [27]. Implementing tiered certification mechanisms, from simulation validation to full-scale flight testing, has proven to be effective in reducing certification cycles by 15%, as seen in the Beta ALIA project which secured preliminary FAA approval. Ultimately, the adoption of transparent AI systems is crucial in ensuring the safety and reliability of intelligent technologies in aviation [28].

5.3 Airspace Management and Air Traffic Control Coordination

The escalating congestion in lower airspace presents a significant hurdle for advanced technologies. AI navigation can be disrupted by real-time traffic, causing conflict probabilities to rise [29]. Traditional air traffic control systems, such as ADS-B, face challenges when incorporating the evolving eVTOL autonomy, resulting in higher operational delays. In a recent report, Europe cautioned that insufficient coordination mechanisms may increase accident rates by five times by 2025, limiting the widespread adoption of electric aircraft. Enhancing coordination and integration among intelligent technologies is crucial to efficiently manage the growing demands of modern airspace.

The advancement of Unmanned Traffic Management (UTM) platforms has become crucial in addressing the challenges of airspace congestion. NASA's UTM 2.0, for example, has achieved an impressive 99.8% conflict avoidance rate through the use of AI multi-agent coordination [30]. By leveraging 5G and satellite links for real-time data sharing, such as in the Vertical VX4 tests where airspace capacity was tripled, system integration is further strengthened. Moving forward, the EU's SESAR Joint Undertaking aims to standardize AI air traffic control interfaces, covering 90% of low-altitude regions [31]. This strategic move is expected to bring efficient collaborative momentum to the global Urban Air Mobility (UAM) ecosystem, paving the way for smoother and safer air travel.

6 Conclusion and Outlook

Electric aircraft represent a pivotal advancement in the realm of air transportation, showcasing significant strides in sustainability and the infusion of intelligent technologies. This comprehensive analysis delves into the pivotal components of electric aircraft, exploring the intricacies of key technologies, applications of intelligent systems, and the current landscape of development trends and challenges. By implementing cutting-edge solutions such as model compression, explainable artificial intelligence (XAI) frameworks, and Unmanned Traffic Management (UTM) platforms, obstacles related to computational resources, safety certifications, and airspace management are effectively mitigated, leading to enhanced autonomy and efficiency within electric aircraft. Projections indicate that by 2030, the integration of intelligent technologies will reach an impressive 85%, a substantial increase from the 40% observed in traditional aircraft. This surge in innovation not only bolsters the sustainability of the aviation industry but also catalyzes transformative changes within the low-altitude economy, revolutionizing short-haul transportation, intercity commuting, and urban combat scenarios. To facilitate the widespread adoption of electric aircraft, the establishment of unified international standards and collaborative efforts across various disciplines are imperative. Future research endeavors are poised to concentrate on hydrogen-electric hybrids and edge AI optimization to explore the vast potential of this "third technological revolution" and deliver groundbreaking solutions to propel global intelligent aviation to new heights.

References

1. Liang X (2020) Research on electric vehicles and their key technologies. Aeronaut Sci Technol 31(6):1–6
2. Zhao C, Chen L, Lu L et al (2012) Advances in electric aircraft technology. Sci Technol Rev 30(12):62–70
3. Li K (2019) Research on the development of electric aircraft technology. Aeronaut Sci Technol 30(1):1–7
4. Qu X (2015) Electric aircraft: leader of the flight revolution. Civ Aircr Manuf 4:40–44
5. Nelson T (2005) 787 systems and performance configured for success. Boeing, Tech Rep. https://www.boeing.com/commercial/787/
6. Wheeler P, Bozhko S (2014) The more electric aircraft: technology and challenges. IEEE Electrif Mag 2(4):6–12
7. Li K (2015) Research on electrical systems of more electric aircraft. Int Aviat 12:26–30
8. Sun X, Cheng W, Mu Z et al (2019) White paper on the development of electric aircraft. Aeronaut Sci Technol 30(11):1–7
9. Peng J (2016) Research on the development of electric aircraft technology. Dual Use Technol Prod 24:30
10. Hoelzen J et al (2021) Energy storage requirements for regional electric aircraft. Prog Aerosp Sci 134:100858
11. Cui Y et al (2023) High-energy and long-cycling all-solid-state batteries enabled by a conductive polymer binder. Nat Energy 8(5):519–528
12. U.S. Department of Energy (2025) Hydrogen and fuel cell technologies office: annual merit review report. https://www.hydrogen.energy.gov/annual-merit-review-2025

13. Battery 2030+ Secretariat (2022) Battery 2030+ roadmap: inventing the sustainable batteries of the future. Available: https://battery2030.eu/roadmap
14. Mouritz AP (2012) Introduction to aerospace materials. Woodhead Publishing, Cambridge
15. Hunston D, Harris C (2017) Advanced composites for aerospace applications: a review of current status and future prospects. NASA technical memorandum TM-2017-219647
16. eVTOL Insights (2025) Global eVTOL market forecast 2025–2030. https://evtolinsights.com/market-forecast-2025
17. Nguyen NV, Kim S (2021) Advanced fault-tolerant control for electric aircraft with distributed propulsion. In: Proceedings of AIAA aviation forum
18. Snyder C, Papanikolopoulos N (2022) Fault-tolerant control challenges and approaches for electric vertical takeoff and landing aircraft. Annu Rev Control 54:332–347
19. ZeroAvia (2025) ZA600 hydrogen-electric propulsion whitepaper. Prod Overview
20. Clean Aviation Joint Undertaking (2021) Strategic research and innovation agenda: a roadmap for achieving sustainability goals in the aviation industry by 2035. European Commission, Brussels
21. Wang J et al (2025) A review and bibliometric analysis of intelligent techniques for battery state estimation in electric aviation. Energy Rep 11:4567–4589
22. IDTechEx: intelligent flight control trends. (2025). https://www.idtechex.com/en/research-report/intelligent-flight-control-2025
23. Beta Technologies (2024) ALIA power allocation optimization. Technical Memorandum, Burlington, VT
24. EHang Holdings (2025) EH216 hub testing and charging optimization. Q2 Oper Rep. https://doi.org/10.2345/ehang-hub-2025
25. European Union Aviation Safety Agency (EASA) (2024) Tiltrotor control testing report. EASA Technical Report no. 2024-001, pp 45–67. https://doi.org/10.2825/12345
26. Lilium GmbH (2025) Jet energy optimization for regional eVTOL. R&D Series 3
27. Aviation Week Network (2025) Predictive Maintenance in electric aircraft: trends and case studies. Aerosp Intell Rep 182(5):34–48. https://doi.org/10.1002/aw-predictive-elec-2025
28. Precedence research (2025) Energy management in eVTOL. https://www.pre-cedenceresearch.com/evtol-energy-management
29. Standards UAM (2025) Federal Aviation Administration (FAA). FAA Advis Circ 90–120:1–150. https://doi.org/10.1234/faa-uam-2025
30. Vertical aerospace (2025) VX4 path prediction algorithms. Navigation R&D Report no. VA-2025-01, 15–32, Bristol, UK. https://doi.org/10.4567/vertical-vx4-path-2025
31. EU Clean Aviation Joint Undertaking (2025) Hybrid mode switching in regional aircraft. Tech Deliv D2.3. https://doi.org/10.2792/67890

A Lung Nodule Detection Method Based on the Improved YOLO11n Model

Dandan Yang and Xiao Wang

Abstract Due to differences in the size, shape, and density of lung nodules, traditional single-scale methods struggle to effectively handle nodules of varying sizes and often face issues of low accuracy and detail loss. This paper proposes a lung nodule detection method based on an improved YOLO11n model. Firstly, in the C2PSA module, the self-attention mechanism is replaced with a multi-frequency multi-scale attention module (MFMSA), which combines frequency-domain and multi-scale features to enhance the detection capability for small targets. Secondly, a modulated fusion module (MFM) replaces the concatenation method in the bottleneck structure, dynamically adjusting weights to improve the representation of key features and the adaptability of the model. Finally, a convolutional additive self-attention mechanism (CASA) is combined with C3K2 to alleviate the loss of detailed features. Experimental results show that the improved YOLO11n-sms model achieves increases of 1.29, 0.71, 0.63, and 1.77 percentage points in precision, recall, mAP50, and mAP50-95, respectively, reaching 92.04, 92.8, 95.5, and 73.0%.

Keyword Lung nodule detection · YOLO11 · Object detection

1 Introduction

With the rapid development of medical imaging technology, especially its applications in computer vision and artificial intelligence, the detection of pulmonary nodules has become a key step in the early diagnosis of lung cancer. As one of the leading causes of cancer-related deaths worldwide, early detection of lung cancer is crucial for improving patient survival rates. Traditional methods for detecting pulmonary nodules largely rely on manual image reading and rule-based image processing algorithms. However, these methods have issues such as long diagnostic

D. Yang · X. Wang (✉)
Guizhou University, Guiyang, Guizhou, China
e-mail: xwang9@gzu.edu.cn

G. Mu et al. (eds.), *Proceedings of the 2nd Electrical Artificial Intelligence Conference, Volume 1*, Lecture Notes in Electrical Engineering 1587,
https://doi.org/10.1007/978-981-95-7932-7_8

times and low accuracy, and they are particularly susceptible to interference when handling complex imaging data, which affects the accuracy of diagnosis.

In recent years, deep learning-based object detection technology has opened up new research avenues for medical image analysis. The YOLO model, due to its efficiency and accuracy, is widely used for automatic detection of pulmonary nodules. Wu and Yali [1] and colleagues proposed a pulmonary nodule detection method based on an improved YOLOv5. By incorporating efficient channel and coordinate attention mechanisms into the backbone and modifying the neck network's feature pyramid into a weighted bidirectional feature pyramid network, they improved the detection model. However, certain limitations remain, as small nodules with inconspicuous features may still be missed. Qiangqiang et al. [2] and colleagues proposed a method to improve the YOLOv8 model by optimizing the C2f module through the RepViTCAA module, enhancing the detection accuracy of tiny pulmonary nodules while achieving a lightweight design. They also introduced the ECLA-HSFPN module to reconstruct the feature fusion part, improving the detection of nodules with varying scales. Additionally, integrating the KAN network improved the detection accuracy of tiny nodules and the model's generalization ability. By using Inner-IoU assisted boxes to enhance the CIoU loss function, the algorithm further addresses issues of varying nodule scales and improves detection accuracy. Nonetheless, the algorithm does not consider the problem of high background noise in pulmonary nodules, leading to decreased performance in noisy conditions. Ji et al. [3] and colleagues proposed the YOLOv5-CASP model, which integrates CBAM, ASPP, and CoT modules to increase pulmonary nodule detection accuracy, but it reduces training and inference speed. Zhu et al. [4] and colleagues proposed an algorithm based on Hierarchical-Split HRNet and a dilated convolution feature pyramid, which uses adversarial training to optimize false positive recognition and improves detection accuracy, albeit with a high computational cost.

YOLO11 has been improved on the basis of previous YOLO versions, incorporating new features and offering fast, accurate, and user-friendly performance, making it the baseline model for this study. Given that pulmonary nodules vary in size and shape and are difficult to detect, this study performed a series of optimizations based on the YOLO11n model, aiming to enhance detection accuracy and computational efficiency. The contributions of this paper are as follows:

(1) Propose replacing the C2PSA module's self-attention mechanism with the MFMSA module, and combine frequency-domain and multi-scale feature extraction methods to enhance small object feature extraction capability and the detection of nodules of different sizes.
(2) By combining the Convolutional Additive Self-Attention (CASA) with C3K2, it can effectively alleviate the potential loss of detailed features in traditional CNNs during image processing.
(3) Introduce the Modulation Fusion Module (MFM) to replace the simple concatenation method in the traditional bottleneck structure, significantly improving the performance of deep learning models in object detection and adaptation to complex scenarios.

2 YOLO11 Object Detection Model

YOLO11 is the latest version in the YOLO series, featuring an improved backbone and neck architecture, which enhances feature extraction capabilities and improves detection accuracy and performance on complex tasks. Its streamlined design and optimized training process achieve faster speeds while maintaining an optimal balance between accuracy and performance. The network structure is shown in Fig. 1. Compared to the YOLOv8 model, YOLO11 replaces the CF2 module with C3K2, adds a C2PSA module after the SPPF module, and introduces the head structure design concept of YOLOv10 into the YOLO11 head module. It also utilizes depth-wise separable convolutions to reduce redundant computations, thereby improving the model's computational efficiency.

YOLO11 introduces the C2PSA module and replaces C2f with C3k2. C3k2 allows for customizable convolution block sizes, providing more flexible configuration. When the hyperparameter C3k = True, the bottleneck block uses C3k. C2PSA extends C2f by incorporating Position-Sensitive Attention (PSA) and Feedforward Neural Networks (FFN), enhancing feature extraction capabilities. It can optionally include a residual structure to optimize gradient propagation and improve training effectiveness. The loss calculation includes two branches: a classification branch

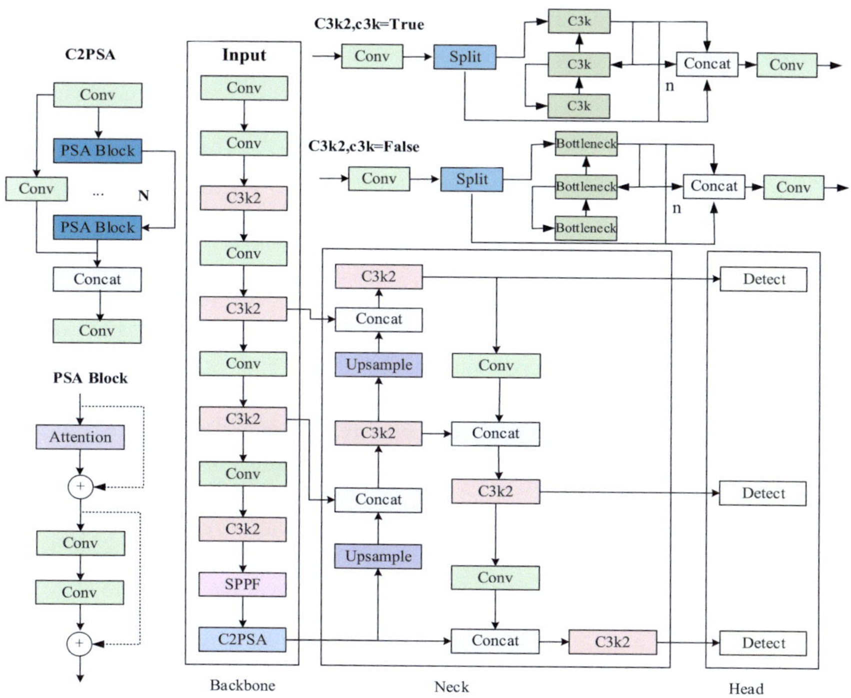

Fig. 1 YOLO11 network architecture diagram

and a regression branch. The classification branch still uses BCE Loss, while the regression branch employs Distribution Focal Loss (DFL Reg_max default is 16) and CIoU Loss.

3 Research Methods

3.1 YOLO11 Model Improvement

This paper addresses issues in lung nodule detection such as large size differences, complex details, and insufficient accuracy of traditional methods, by proposing a detection method based on an improved YOLO11n model. The network structure of the improved model is shown in Fig. 2. Firstly, the MFMSA module is used to replace the self-attention mechanism of C2PSA, combining frequency domain and multi-scale analysis to enhance the feature extraction capability for small targets and multi-scale features. Secondly, the MFM (Modulated Fusion Module) replaces the bottleneck structure's contact, dynamically adjusting weights to highlight key features and enhance feature interaction and representation capabilities. Finally, a Convolutional Additive Self-Attention mechanism (CASA) is introduced and combined with C3K2, utilizing global information to mitigate detail loss in CNNs and improve adaptability in complex scenarios.

3.2 MFMSA Module

The MFMSA module effectively adapts to pulmonary nodules of different sizes and details through multi-scale feature extraction and multi-frequency attention mechanisms, reducing noise interference and enhancing the model's perception of nodule features, thereby improving detection accuracy. The structure of MFMSA mainly includes the following key components, as shown in Fig. 3. Scale Decomposition Module: The MFMSA module employs a multi-branch architecture to extract multi-scale features. Each branch processes the input feature map, focusing on different scales of image features by reducing the resolution and number of channels. Branches at different scales focus respectively on large-scale structural information and small-scale detailed features, thus forming complementary multi-scale data support. To generate multi-scale features from an input feature map Xi, assume there are S branches operating at S different scales. For each branch at the s-th scale ($1 \leq s \leq S$), the resolution and channels of the input feature map Xi are reduced using a ratio $\gamma \in (0, 1)$ to improve computational efficiency. Each input feature map at the s-th scale branch can be expressed as follows:

$$X_i^s = \mathrm{Conv2D}_{2s+1}(\mathrm{Down}_s(X_i)) \in R^{C_s \times H_s \times W_s} \tag{1}$$

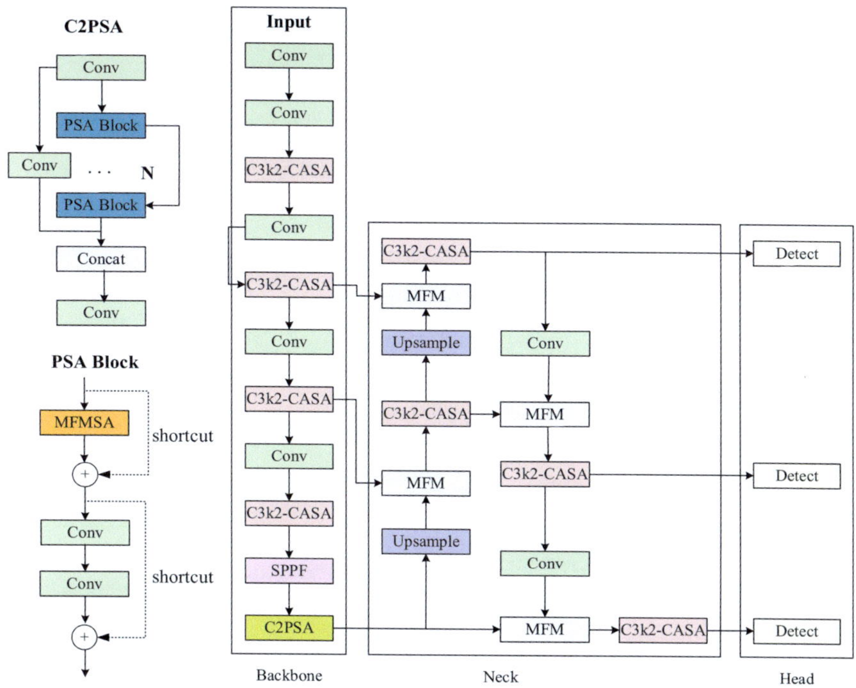

Fig. 2 Improved model network structure diagram

Among them, Downs(·) is a downsampling with a scaling factor of 2^{s-1}. In addition, the number of channels, height, and width of the s-th scale branch are represented as $C_s = \max\left(\frac{C}{\gamma^{s-1}}, C_{min}\right)$, $H_s = \max\left(\frac{H}{2^{s-1}}, H_{min}\right)$, $W_s = \max\left(\frac{C}{2^{s-1}}, W_{min}\right)$.

Multi-Frequency Channel Attention Module: This module extracts frequency-domain features based on 2D Discrete Cosine Transform (2D DCT) techniques [5–7]. After processing the features of each scale branch, the features are compressed by applying global average pooling, global max pooling, and global min pooling operations, and then frequency statistics are aggregated through two fully connected layers, generating a channel attention map. This attention map adjusts the original feature map, dynamically enhancing critical channel information while suppressing noise interference, thereby improving feature discriminability. The feature representation at each scale branch is as follows:

$$X_i^{s,k} = \sum_{h=0}^{H_s-1} \sum_{w=0}^{W_s-1} \left(X_i^s\right)_{:,h,w} D_{h,w}^{u_k,v_k} \tag{2}$$

Here, (u_k, v_k) are the frequency indices of $X_i^{s,k}$ and $D_{h,w}^{u_k,v_k} = \cos\left(\frac{\pi h}{H_s}\left(u_k + \frac{1}{2}\right)\right)\cos\left(\frac{\pi w}{W_s}\left(v_k + \frac{1}{2}\right)\right)$.

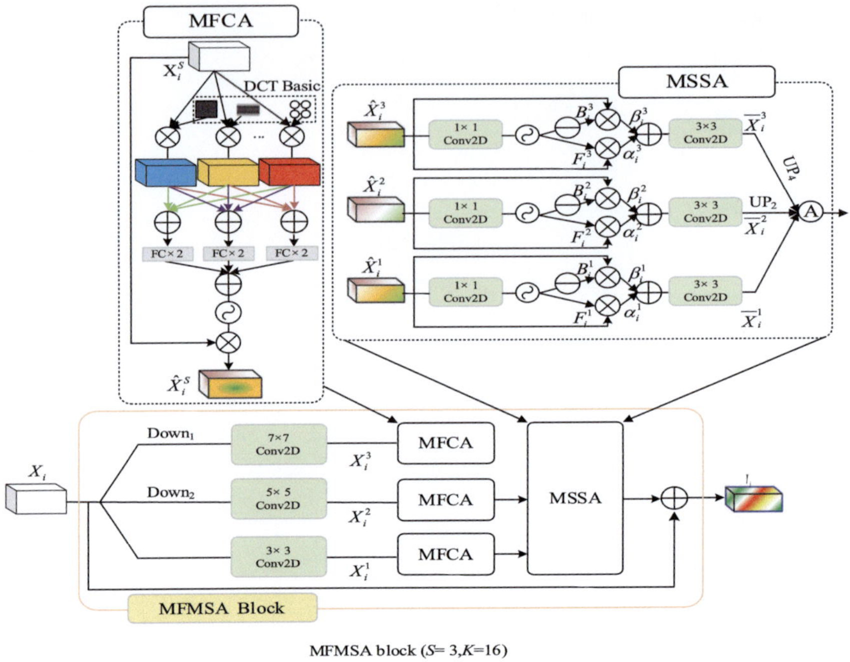

Fig. 3 MFMSA module

Multi-scale spatial attention module: Based on the feature maps after channel recalibration, it further explores discriminative boundary cues at different scales in the spatial domain. Learnable parameters are introduced for each scale branch to separately control the information flow of the foreground and background. After computing the feature maps with spatial attention, the features from different scale branches are aggregated through upsampling operations, and residual connections are added, enabling the model to better capture boundary features while reducing noise interference and enhancing the ability to identify lesion boundaries.

3.3 MFM Module

In YOLOv11, the Modulation Fusion Module (MFM) replaces concatenation, allowing dynamic adjustment of weights based on feature importance. Compared to simple concatenation, it more effectively emphasizes key features, enhances feature interaction, and improves the model's ability to represent target features. The structure is shown in Fig. 4b. The core principle of MFM is based on feature fusion [8] and weight modulation [9]; by integrating features from different sources and dynamically adjusting the weights, it highlights features that aid in dehazing and suppresses distracting features, thereby enhancing the image feature representation and optimizing dehazing performance. It can adaptively integrate information from different scales and types, helping the network learn clear image structures and details, improving detection accuracy and adaptability in complex scenarios.

The Modulation Fusion Module (MFM) mainly includes several key steps: first, it receives feature information from other modules, including features obtained from embedding local features into the global feature extraction module and features processed through convolution. Next, global average pooling is used to convert the input feature maps into a vector representing global information, allowing an understanding of the overall context of the features. Then, this global information vector is processed through a multilayer perceptron, and a coefficient matrix is generated using the Softmax function to assess the importance of each feature. Based on this coefficient matrix, the input features are weighted, emphasizing important features and fusing them together. Finally, the fused features are output after convolution and other operations, ready for use by subsequent network modules.

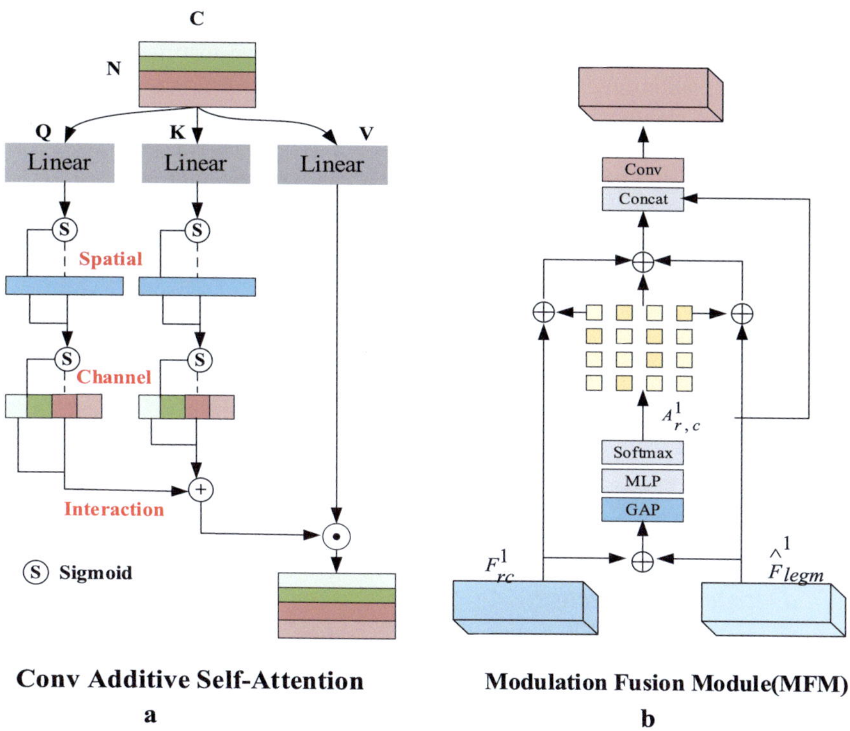

Fig. 4 CASA module, MFM module

3.4 CASA Module

The Convolutional Additive Self-Attention (CASA) mechanism reduces computational complexity by simplifying the similarity function, defining it as the sum of the contextual scores between Query (Q) and Key (K). Contextual scores are calculated using a mapping function that combines spatial and channel attention, allowing the network to flexibly learn information correlations. Multiple interactions between Q and K help the network focus on key information tokens, reducing computation while retaining important information, balancing performance and efficiency. The CASA module relies on the Convolutional Additive Token Mixer (CATM) to process input tensors in both spatial and channel domains. The CASA module mainly consists of the following components:

Spatial Domain Operations: A 3 × 3 convolutional layer is first used to extract features from local regions. This layer has an appropriate receptive field, effectively integrating local information from the input image. Then, a 1 × 1 convolutional layer reduces the number of feature channels to 1 while keeping the spatial dimensions unchanged, generating a spatial attention map. Finally, the map is processed by a Sigmoid activation function, mapping output values between 0 and 1 [10]. The resulting spatial attention map reflects the importance of each position, with larger values indicating higher importance.

Channel Domain Operations: The channel domain operation draws inspiration from SENet [11] (Squeeze-and-Excitation Networks) but differs in approach. SENet integrates inter-channel information through channel reduction, whereas the Conv Additive Self-Attention directly consolidates inter-channel information using a 1 × 1 convolutional layer. The 1 × 1 convolution can perform weighted summation of each channel's information, achieving inter-channel interaction. After processing with the Sigmoid activation function, a channel attention map [12] is obtained, reflecting the importance of different channels.

4 Experimental Section

4.1 Dataset Introduction

This study uses the LUNA16 public dataset, which contains 888 cases and 36,378 lung nodules. Nodules larger than 3 mm were selected as the subjects of this study, among which 1,186 were confirmed positive nodules. Nodule locations and information were annotated by four professional doctors. The dataset was split into training, testing, and validation sets at a ratio of 7:1.5:1.5.

4.2 Experimental Setup and Parameter Settings

The experimental environment is based on the Windows 11 operating system, with an Intel Core i7-14700K processor running at 3.4 GHz, using a 64-bit architecture and equipped with 32 GB of memory. The software environment uses the PyTorch-CUDA deep learning framework for development and computation, with version PyTorch 12.4. The programming tool used is PyCharm 2022, and the programming language is Python 3.12. Regarding hyperparameter settings, the total number of iterations is 200 epochs, the input image size is 640 × 640, the batch size is 16, the initial learning rate is set to 0.01, and the optimizer chosen is SGD.

4.3 Test Results and Analysis

During the training process of the improved YOLO 11n model, the Box Loss and Cls Loss values, as well as the model's precision, recall, and mAP50, were recorded. The change curves of loss values, precision, recall, and mAP with the number of iterations during training are shown in Fig. 5. In the early stages of training, the loss values decreased rapidly, but by the 20th epoch, the rate of decline had significantly slowed, eventually stabilizing around the 200th epoch. At the 200th epoch, the Box Loss was 0.9352, and the Cls Loss was 0.5695, indicating that the model had basically converged. After 200 epochs of training, the model achieved 92.04% precision, 92.82% recall, and 95.48% mAP.

4.4 Evaluation Metrics

This study uses Precision, Recall, and mean Average Precision (mAP50, mAP50-95) to measure model accuracy, and model weights to measure model complexity. Specifically:

Precision: The proportion of true positive samples among all samples predicted as positive by the model.

Recall: The proportion of true positive samples correctly predicted by the model among all actual positive samples.

mAP50: The average precision at an IoU threshold of 0.5.

mAP50-95: The average precision with the IoU threshold gradually varying from 0.5 to 0.95, providing a more comprehensive evaluation of accuracy.

The formulas are as follows:

$$P = \frac{TP}{TP + FP} \tag{3}$$

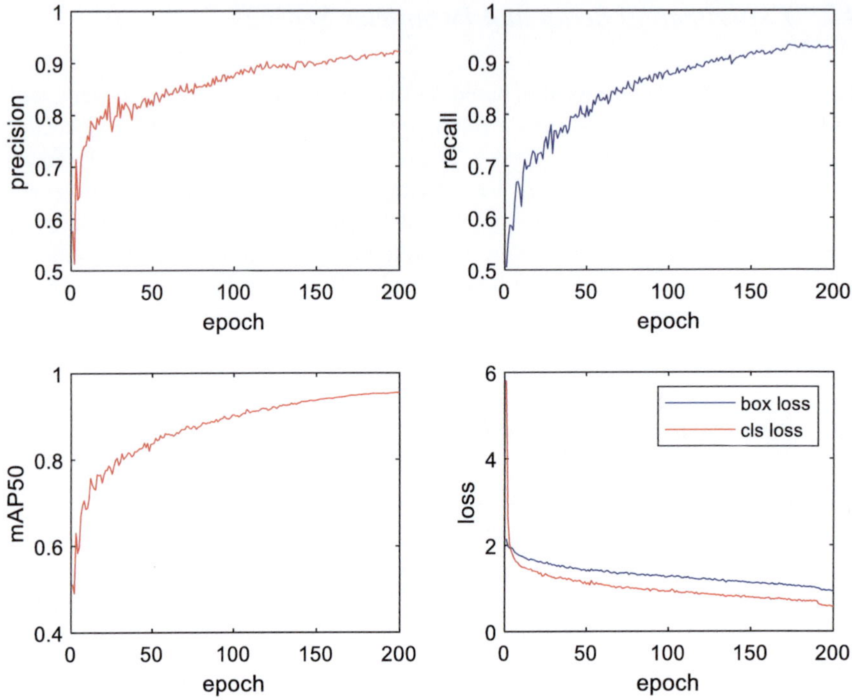

Fig. 5 Curves of precision, recall, mAP, and loss values

$$R = \frac{TP}{TP + FN} \tag{4}$$

$$mAP50 = \frac{1}{N}\sum_{i=1}^{N} AP_i \tag{5}$$

$$mAP50\_95 = \frac{1}{N}\sum_{i=1}^{N}\frac{1}{10}\sum_{t=50}^{95} AP_t(i) \tag{6}$$

Here, TP refers to the samples correctly predicted as positive, FP refers to the samples incorrectly predicted as positive, and FN refers to the samples incorrectly predicted as negative. AP_i is the average precision of the i-th category, with the IoU threshold set to 0.5. $AP_t(i)$ represents the average precision of the i-th category at a specific IoU threshold.

Table 1 Ablation experiment results (%)

Model	Precision	Recall	mAP_{50}	mAP_{50-95}
YOLO11n	90.75	92.11	94.85	71.23
YOLO11n+M1	91.19	92.98	94.96	71.42
YOLO11n+M2	91.26	92.77	95.03	71.61
YOLO11n+M3	91.17	92.48	94.84	71.11
YOLO11n+M4	91.48	92.72	95.07	71.70
YOLO11n+M5	91.45	93.03	95.14	71.96
YOLO11n+M6	91.89	92.60	95.24	72.28
YOLO11n+M7	92.04	92.82	95.48	73.00

4.5 Ablation Experiments

Using YOLO 11n as the baseline model, eight sets of ablation experiments were conducted under the same training environment and hyperparameter settings to verify the effectiveness of each module in the improved YOLO 11 network model. Here, M1 represents MFMSA, M2 represents CASA, M3 represents MFM, M4 represents MFMSA CASA, M5 represents MFMSA MFM, M6 represents CASA MFM, and M7 represents MFMSA CASA MFM. The results are shown in Table 1.

4.6 Comparison Experiment with Mainstream Models

In order to verify the advantages of the algorithm proposed in this paper for lung nodule detection, the model in this paper was compared with several mainstream models. The experimental dataset and parameter settings were consistent, and the results are shown in Table 2.

Table 2 Comparison of detection performance of different models (%)

Model	Precision	Recall	mAP_{50}	mAP_{50-95}	Weight/MB
SSD	88.2	85.2	89.9	56.5	23.9
YOLOv5	88.9	90.4	92.7	66.5	3.78
YOLOv6	89.9	89.5	93.0	67.0	9.54
YOLOv7	88.6	91.2	92.8	66.9	71.3
YOLOv8	90.2	91.7	93.6	68.4	15.49
YOLOv9	90.9	92.0	94.5	70.5	23.7
YOLOv10n	90.8	92.2	93.8	70.3	34.7
Our algorithm	92.0	92.8	95.5	73.0	10.4

SSD [13] exhibited stability, but its precision and recall were relatively low; YOLOv5 [14] and YOLOv7 [15] showed similar performance, whereas YOLOv6 [16] achieved better results in Precision and mAP50. For YOLOv8 [17] and YOLOv9, detection performance improved significantly, with all metrics surpassing previous generations, and YOLOv9 [18] outperformed YOLOv8 in detection accuracy. For YOLOv10 [19], all metrics slightly declined compared to YOLOv9. The algorithm proposed in this paper exceeds the comparative models in terms of precision, recall, and mean average precision, especially showing notable improvements in recall and mAP. A higher recall (92.8%) can reduce missed detections, making it more suitable for lung nodule screening. A superior mAP50-95 (73.0%) indicates stronger capability in localizing nodule boundaries, aiding subsequent medical analysis. Overall, for lung nodule detection on the LUNA16 dataset, the proposed algorithm outperforms several YOLO series models and traditional models, while maintaining lightweight model weights, fully demonstrating the effectiveness of the improvements made.

4.7 *Visualization of Detection Results*

Based on the experimental results of the improved YOLO11n model, a visualization analysis was conducted. As shown in Figs. 6, 7, and 8, blue rectangular boxes indicate the model's predicted bounding boxes, displaying lung CT images with clearly marked nodule positions (labeled as "nodule"). The YOLO11n model successfully detected lung nodules and provided confidence scores, such as "nodule 0.74", reflecting the model's detection accuracy at that location. The improved YOLO11n model not only accurately localized the nodules but also increased the confidence scores, indicating a significant enhancement in performance for lung nodule detection.

By comparing the original annotations, YOLO11n, and the improved model's detection results, the significant improvement in lung nodule detection is demonstrated, proving that the proposed algorithm outperforms YOLO11n and other mainstream models. The proposed improvement exhibits higher accuracy and reliability, confirming the advantages of this model in lung nodule detection.

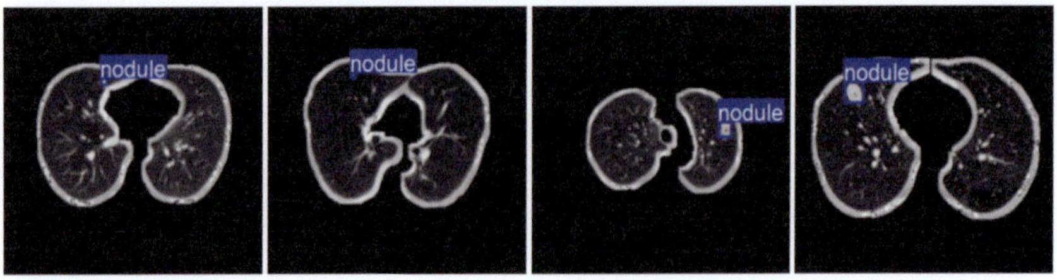

Fig. 6 Lung nodule annotation

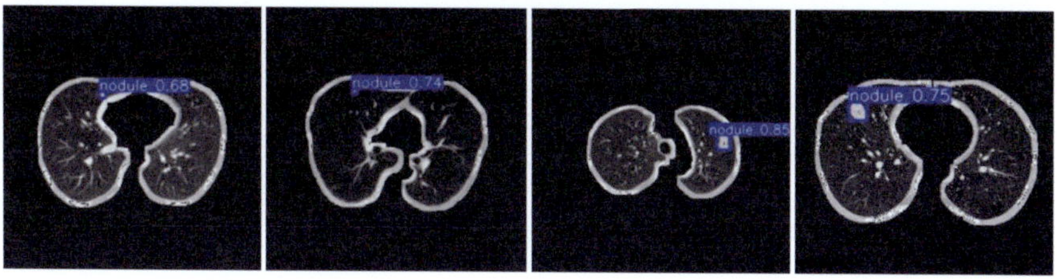

Fig. 7 YOLO11n detection results

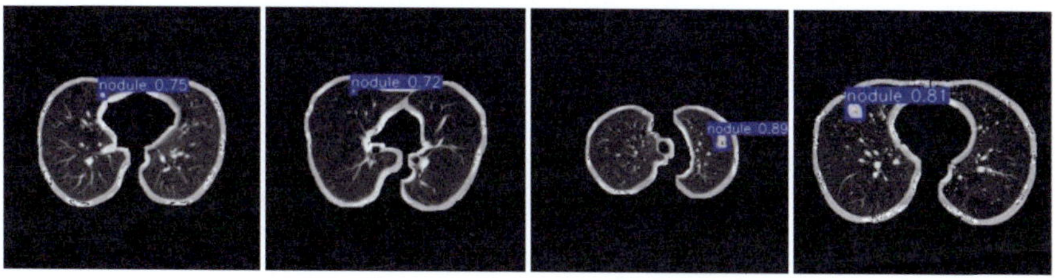

Fig. 8 Improve YOLO11n detection results

5 Conclusion

This paper proposes an improved YOLO11 object detection method, which significantly enhances the extraction of small object features and the detection of nodules of different sizes by introducing the MFMSA module to replace the traditional C2PSA module and combining frequency domain and multi-scale feature extraction. The method integrates the Convolutional Additive Self-Attention (CASA) mechanism and the C3K2 module, alleviating the loss of detailed features caused by traditional CNNs, thereby improving detection accuracy and detail restoration capability. The introduction of the Modulation Fusion Module (MFM), replacing the traditional concatenation approach, further enhances the model's performance in complex scenarios. After 200 training epochs, the model achieved a precision of 92.04%, a recall of 92.82%, and an mAP of 95.48%, significantly outperforming YOLO11n, particularly in recall and mAP50-90 (73.0%), indicating that this model has stronger detection capability and accuracy for lung nodule screening.

References

1. Wu C, Yali L (2024) An improved YOLOv5-based method for lung nodule detection in CT images. Comput Sci 51(S1):587–592
2. Qiangqiang B, Siyuan T, Li, Qingqian et al (2025) Improved YOLOv8 model for multi-type lung nodule detection. Comput Sci Explor 19(02):429–444

3. Ji ZL, Wu Y, Zeng XY et al (2023) Lung nodule detection in medical images based on improved YOLOv5s. IEEE Access 11:76371–76387
4. Zhu L, Zhu H, Yang S et al (2023) Pulmonary nodule detection based on hierarchical-split HRNet and feature pyramid network with atrous convolution. Biomed Signal Process Control 85:105024
5. Gu AR, Nam JH, Lee SC (2022) Fbi-net: frequency-based image forgery localization via multitask learning with self-attention. IEEE Access 10:62751–62762
6. Qin Z, Zhang P, Wu F, Li X (2021) Fcanet: Frequency channel attention networks. In: Proceedings of the IEEE/CVF international conference on computer vision, pp 783–792
7. Sang M, Hansen JHL (2022) Multi-frequency information enhanced channel attention module for speaker representation learning. arXiv preprint arXiv:2207.04540
8. Zhuo D, Hesheng Z, Yicong T et al (2025) Diagnosis method for broken bars in traction motor rotors based on convolutional neural networks with multi-scale feature fusion [J/OL]. Trans Chin Inst Electr Eng 1–15
9. Zhu Z, Zhang L, Pei K et al (2025) Endpoint-aware audio-visual speech enhancement utilizing dynamic weight modulation based on SNR estimation. Neural Netw 185:107152–107152
10. Jiang S, Yu S (2025) Optimizing multi-focus image fusion through convolutional attention vision transformers and spatial consistency models. Appl Soft Comput 181 113507–113507
11. Hu J, Shen L, Sun G (2018) Squeeze-and-excitation networks. In: Proceedings of the IEEE conference on computer vision and pattern recognition, pp 7132–7141
12. Zhang Y, Su C, He X et al (2025) Progressive hybrid hypergraph attention network with channel information fusion for remaining useful life prediction of rolling bearings. Mech Syst Signal Process 236 112987–112987
13. Liu W, Anguelov D, Erhan D et al (2016) SSD: single shot MultiBox detector. In: Proceedings of the 14th 284 European conference on computer vision, pp 21–37
14. Vidhya NP, Priya R (2024) Automated diagnosis ofthe severity of TMB infestation in cashew plantsusing YOLOv5. J Image Graph 12(3):276–282
15. Wang CY, Bochkovskiy A, Liao HYM (2023) YOLOv7: trainable bag-of-freebies set new state-of-the-art for real-time object detectors. In: 2023 IEEE/CVF conference on computer vision and pattern recognition (CVPR). IEEE Press, Vancouver, Canada, pp 7464–7475
16. Li C, Li L, Jiang H et al (2022) YOLOv6: a single-stage objectdetection framework for industrial applications. arXivpreprint arXiv:2209.02976
17. Liu Z, Zhong X, Wang C, Wu G, He F, Wang J, Yang D (2025) Rapid and accurate detectionof peanut pod appearance quality based onlightweight and improved YOLOv5_SSE model. Front Plant Sci 16:1494688
18. Zhoujx LZ, Hao YJ (2024) Method for defect detection on steel plate surface based on improved YOLOv9. Electron Meas Technol 47(22):181–188
19. Liu R (2024) Enhanced motorbike helmet detection usingYOLOv10 with attention-based modifications. In: 4th international conference on computer science and blockchain (CCSB). IEEE, 114–118
20. Ji X, Chenrong H, Junci Y et al (2025) Research on insulator detection combining branch characteristics and rejection loss. Comput Eng Appl 61(07):141–152

Effects of Alternating Current Stimulation on the Firing Parameters of Excitatory and Inhibitory Neurons

Yifan Li, Changda Liu, Yujie Zhao, Haoyu Qiu, Pengfei Zhao, and Haijun Zhu

Abstract Transcranial alternating current stimulation (tACS), as a non-invasive neuromodulation technique, is widely used in fields such as neuroscience research, making it an important tool for exploring the working mechanisms of the nervous system and understanding brain functions. Based on the Izhikevich neuron model, this paper investigates the response characteristics of excitatory and inhibitory neurons under varying stimulation intensities of alternating current. The results indicate that the stimulation intensity significantly influences neuronal firing behavior: as the stimulation intensity increases, the firing frequency of both types of neurons increases, the half-peak width shortens, and the peak amplitude of action potentials increases after the stimulation intensity reaches a certain level. Meanwhile, the membrane potential amplitude first rises and then declines. These findings reveal that transcranial alternating current stimulation can modulate neuronal excitability by adjusting the stimulation intensity, providing theoretical guidance for the practical application of alternating current stimulation.

Keywords Transcranial alternating current stimulation · Izhikevich neuron model · Excitatory neurons · Inhibitory neurons

Y. Li · C. Liu · Y. Zhao · H. Qiu · H. Zhu (✉)
Key Laboratory of Digital Medical Engineering of Hebei Province, College of Electronic and Information Engineering, Hebei University, Baoding, Hebei, P.R. China
e-mail: suseyinxiang@126.com

Y. Zhao
e-mail: 17796971534@126.com

H. Qiu
e-mail: 15737123022@126.com

P. Zhao (✉)
Department of Neurosurgery, Affiliated Hospital of Hebei University, Baoding, Hebei, P.R. China
e-mail: zpf03539@126.com

G. Mu et al. (eds.), *Proceedings of the 2nd Electrical Artificial Intelligence Conference, Volume 1*, Lecture Notes in Electrical Engineering 1587,
https://doi.org/10.1007/978-981-95-7932-7_9

1 Introduction

The firing activity of neurons is the foundation of information encoding and transmission in the nervous system. Its firing patterns are regulated by various factors, such as ion channel properties, synaptic connections, and external stimuli. Different types of neurons, such as excitatory neurons and inhibitory neurons, play distinct roles in neural networks, collectively maintaining the balance and function of neural circuits. Excitatory neurons primarily promote the firing of downstream neurons by releasing excitatory neurotransmitters like glutamate, while inhibitory neurons regulate network activity through inhibitory neurotransmitters such as GABA, preventing excessive excitation [1, 2].

Transcranial alternating current stimulation (tACS), as an emerging non-invasive brain stimulation technique, plays a significant role in modulating cortical neural activity and synaptic plasticity, as well as improving cognitive functions in neurological and psychiatric disorders. tACS primarily operates by applying low-intensity alternating sinusoidal currents at specific frequencies to entrain cortical oscillations, thereby regulating brain activity [3]. Compared to conventional non-invasive neuromodulation techniques such as transcranial magnetic stimulation (TMS) and transcranial direct current stimulation (tDCS), tACS offers distinct advantages including enhanced safety, lower cost, and superior noise resistance. Furthermore, it enables the elucidation of causal mechanisms underlying neurocognitive processes through frequency-specific entrainment of neural oscillations [4]. In investigations of tACS modulation effects, stimulation frequency plays a critical role. tACS administered at specific frequencies has been demonstrated to modulate diverse sensory and cognitive functions [5]. Research by Norbert et al. [6] provides evidence that theta-frequency band stimulation contributes to the facilitation of memory consolidation processes. Research by Müller [7] reveals that the application of alpha-tACS to the primary visual cortex significantly enhances performance in visual orientation discrimination tasks. These observations indicate that the enhancement of alpha-band oscillations induced by alpha-tACS in the primary visual cortex is mechanistically linked to improved discrimination capability for orientation differences in visual stimuli. Furthermore, additional studies have demonstrated that gamma-tACS and theta-tACS stimulation can enhance motor learning and working memory capabilities [8]. These findings further substantiate the modulatory effects of frequency-specific tACS on cognitive functions and provide emerging evidence for its potential in cognitive enhancement and clinical applications. However, the mechanistic underpinnings of tACS, particularly its precise effects on the firing patterns of distinct neuronal subtypes, remain inadequately explored and require systematic theoretical investigation.

Current research has begun to explore the effects of alternating current stimulation on the firing behavior of Izhikevich (IZH) neuronal models. For instance, tACS at specific frequencies can induce resonance phenomena, thereby enhancing neuronal responsiveness, while variations in stimulation intensity may trigger transitions in firing patterns [9, 10]. However, existing studies have predominantly focused on single neuronal types, lacking a systematic comparison of the response differences between excitatory and inhibitory neurons under tACS stimulation.

2 Methods

2.1 Improved Euler Method for Solving the Izhikevich Neuronal Model

The Izhikevich (IZH) spiking neuron model, proposed by Izhikevich in the early twenty-first century, was developed to reproduce known spiking patterns and bursting behaviors by combining continuous spike-generation mechanisms with discontinuous post-spike reset processes [11]. The Izhikevich (IZH) model offers significant advantages in practical applications due to its concise structure and low computational cost. Crucially, it closely captures key characteristics of biological neurons, including excitability, inhibitory properties, and synaptic transmission mechanisms. These attributes have established the IZH model as a widely recognized tool in computational neuroscience research. The Izhikevich (IZH) model is defined by the following two coupled differential equations:

$$\frac{dv}{dt} = 0.04 \times v^2 + 5 \times v + 140 - u + I \tag{1}$$

$$\frac{du}{dt} = a(bv - u) \tag{2}$$

If v $\geq$ 30 mV, then v = c, u = u + d. Among these, v represents the neuronal membrane potential, u represents the recovery variable of the membrane potential after firing a spike, and I represents the input current. The dimensionless parameters a, b, c, and d are defined as follows: Parameter a describes the time scale of the recovery variable u–a smaller value indicates slower recovery. Parameter b describes the sensitivity of the recovery variable u to the membrane potential v–a larger value leads to stronger subthreshold oscillations and a lower threshold for spike generation. Parameter c describes the post-spike reset value of the membrane potential v, caused by fast high-threshold K^+ conductance. Parameter d describes the post-spike reset value of the recovery variable u, caused by slow high-threshold Na^+ and K^+ conductance.

In the Izhikevich model, considering that the resting membrane potential of neurons ranges from −70 to −60 mV, the resting membrane potential is set to v = −65 mV. The fixed parameters for six types of excitatory cortical spiking neuron models and inhibitory cortical spiking neuron models are listed in Table 1. Numerical simulations are conducted using simulation software, employing an improved Euler method to compute the value of variable v at each time step dt. The equations are as follows:

$$K_1\_v = dt(0.04 \times v^2 + 5 \times 140 - u + I) \tag{3}$$

$$K_1\_u = dt \times a(bv - u) \tag{4}$$

$$K_2\_v = dt[0.04 \times (v + K_1\_v)^2 + 5 \times (v + K_1\_v) + 140 - K_1\_u + I] \tag{5}$$

$$K_2\_u = dt \times a \times [b \times (v + K_1\_v)] - (u + K_1\_u) \tag{6}$$

The predicted value for the n + 1 step is calculated as follows:

$$v(n + 1) = v(n) + 0.5 \times (K_1\_v + K_2\_v) \tag{7}$$

$$u(n + 1) = u(n) + 0.5 \times (K_1\_u + K_2\_u) \tag{8}$$

These equations indicate that the core idea of the improved Euler method is to more accurately predict the future states of the neuronal membrane potential and recovery variable through a prediction step (K1_v and K1_u) and a correction step (K2_v and K2_u).In the Izhikevich model, the recovery variable plays a critical role in simulating the slow internal processes of neurons, which typically involve the activation and inactivation of ion channels. By updating the value of the recovery variable using the improved Euler method, the model can more accurately capture the dynamic influence of these slow processes on the membrane potential, thereby more realistically reflecting the behavior of neurons during electrophysiological activity. Compared to the simple Euler method, the improved Euler method enhances simulation accuracy through its predictor–corrector steps. This approach helps more precisely simulate neuronal responses to stimuli. Additionally, the improved Euler method demonstrates strong numerical stability when handling complex nonlinear differential equations like those in the Izhikevich model, which is essential for simulating highly nonlinear systems such as neurons.

Table 1 Dimensionless parameters of the six neuron models

Neuron types		a	b	c	d
Excitatory neurons	RS	0.02	0.20	−65	8
	IB	0.02	0.20	−55	4
	CH	0.02	0.20	−50	2
Inhibitory neurons	FS	0.10	0.20	−65	2
	LTS	0.02	0.25	−65	2
	LS	0.10	0.20	−65	2

2.2 The Significance and Calculation of Action Potential Characteristic Parameters

The half-peak width (HPW) reflects the overall duration of the action potential. The calculation formula is as follows:

$$HPW = t_2 - t_1 \tag{9}$$

Among these, t_1 and t_2 correspond to the two time points at which the ascending and descending branches of the action potential pass through 50% of the peak amplitude.

The firing rate (FR) reflects the frequency at which a neuron generates action potentials per unit of time. The calculation formula is as follows:

$$FR = \frac{N}{T} \tag{10}$$

where N represents the number of action potentials within a given time window T.

The action potential peak (AP-Peak) reflects the maximum degree of depolarization reached during an action potential. The calculation formula is as follows:

$$AP - Peak = V_{peak} - V_{baseline} \tag{11}$$

where V_{peak} represents the peak voltage, and $V_{baseline}$ denotes the resting membrane potential.

The action potential amplitude (AMP) reflects the maximum voltage deviation of the membrane potential relative to a baseline level (typically the resting membrane potential). The calculation formula is as follows:

$$AMP = V_{peak} - V_{after\ hyper\ polarization} \tag{12}$$

where $V_{after\ hyper\ polarization}$ represents the voltage of the subsequent hyperpolarization trough.

3 Result

3.1 The Effects of Alternating Current Stimulation on the Firing Patterns of Excitatory Neurons

To investigate the effects of alternating current stimulation on the firing frequency of excitatory neurons, simulations based on the Izhikevich (IZH) model were conducted with other input parameters set as fixed values. The changes in membrane potential and firing frequency were simulated within a stimulation intensity range of 2–45 pA. Figure 1 displays the action potential firing patterns of three types of excitatory neurons under stimulation intensities of 2, 4, 6, 8, 10, 15, 20, 25, 30, 35, 40, and 45 pA. It can be observed that the firing frequency of all three types of excitatory neurons increases with the stimulation intensity. Among them:

RS (Regular Spiking) neurons are the most sensitive to sinusoidal current, exhibiting sparse and irregular firing at low currents and transitioning to high-frequency periodic firing at high currents.

IB (Intrinsically Bursting) neurons maintain stable firing patterns throughout. They display a double-peak pattern at 15 pA, show regular periodic firing between 20–40 pA, and maintain high-frequency stable firing even at high currents.

CH (Chattering) neurons initially fire in potential bursts within the 4–20 pA range. Beyond 25–45 pA, the bursting pattern disappears and transitions into high-frequency regular firing.

Figure 2a shows the changes in the half-peak width (HPW) of three types of excitatory neurons (RS, IB, CH) under sinusoidal current stimuli of different intensities. As the stimulation intensity increased, the HPW of all three neuron types exhibited a decreasing trend, indicating that action potentials became faster and sharper under strong stimulation. The phenomenon may be related to the rapid activation and inactivation mechanisms of ion channels.

Figure 2b illustrates the variation in firing rate (FR) with increasing stimulation intensity. The FR of all three neuron types rose with stronger stimuli, demonstrating a positive correlation between stimulation intensity and discharge frequency. The FR of RS neurons tended to stabilize at higher stimulation intensities, while the FR of IB and CH neurons continued to increase even under high-intensity stimulation. The suggests that as stimulation intensity grows, neuronal firing frequency accelerates, with IB and CH neurons exhibiting a more sustained response to high intensities.

Figure 2c depicts changes in the action potential peak (AP-Peak) of the three excitatory neuron types. After an initial increase at lower stimulation intensities, the AP-Peak remained stable and showed no significant changes with further increases in sinusoidal current intensity.

Figure 2d reflects changes in action potential amplitude (AMP). Beyond 20 pA, the AMP of all three neuron types decreased gradually with increasing stimulation intensity.

Under sinusoidal electrical stimulation, all three types of excitatory neurons exhibited adaptive adjustments in response to the stimulation intensity. The decrease in HPW indicates that action potentials become faster and sharper under strong stimulation, which may be associated with rapid kinetic changes in ion channels. The increase in FR reflects the elevated firing frequency of neurons under intense stimulation, representing a typical neuronal response to strong stimuli. This reductions in AP-Peak and AMP may reflect adaptive mechanisms of neurons under sustained strong stimulation, such as depolarization block effects and sustained depolarization.

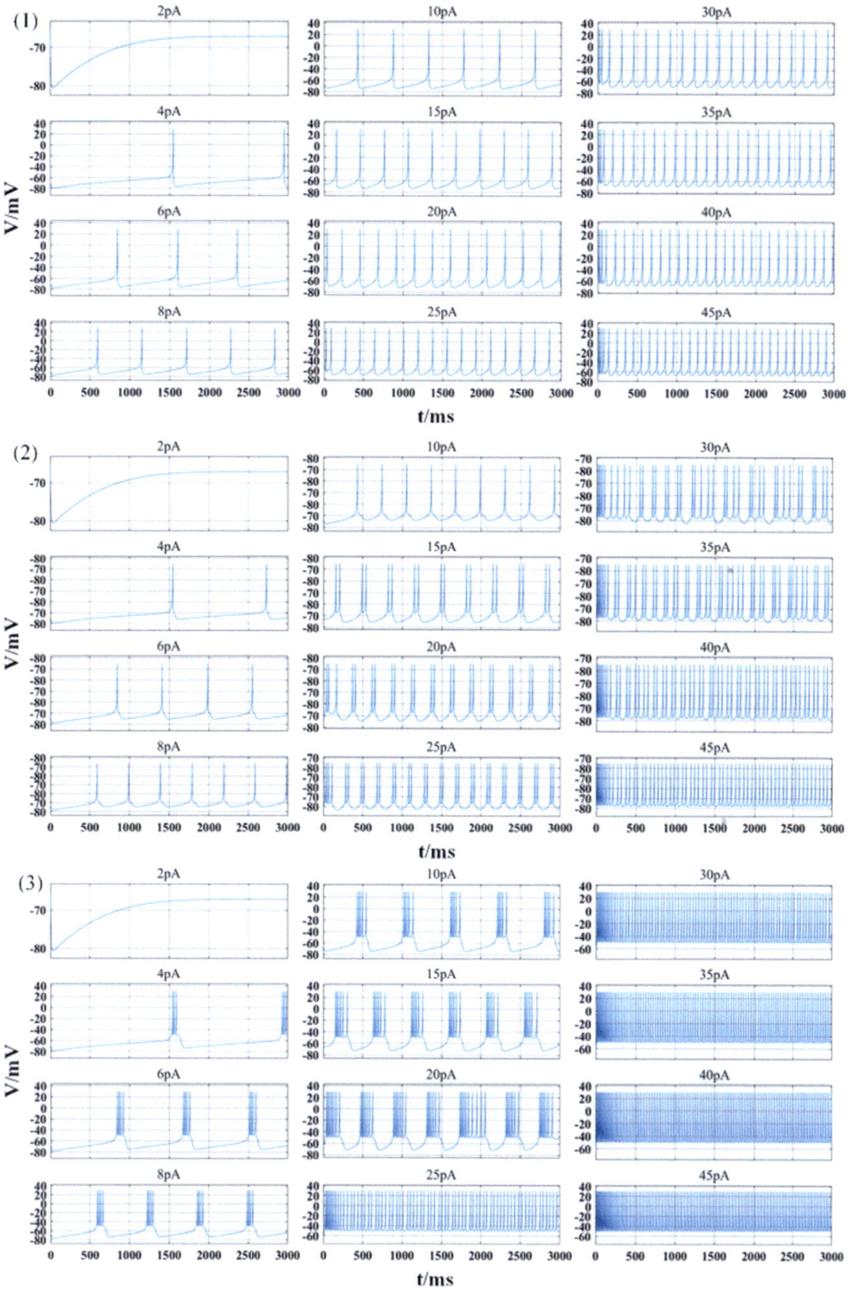

Fig. 1 Simulation results of action potentials in excitatory neurons under sinusoidal electrical stimulation

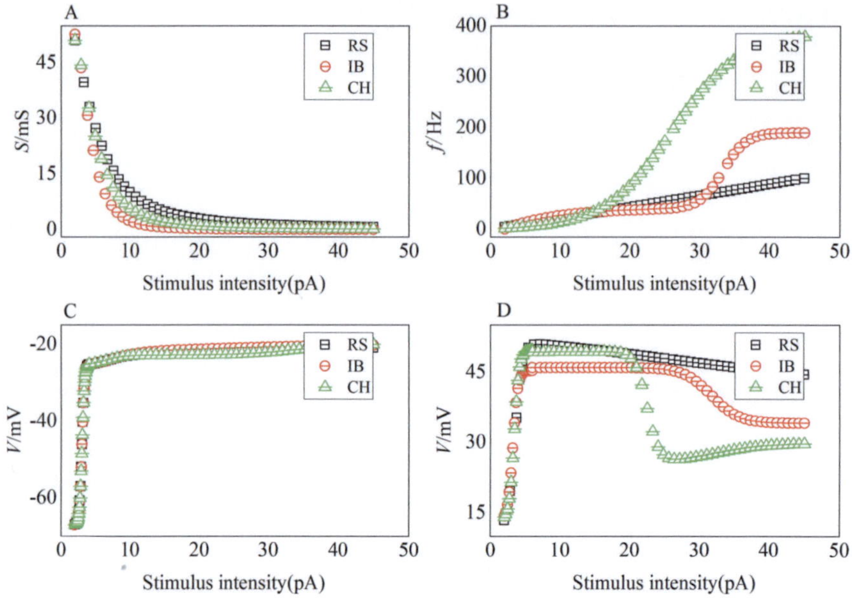

Fig. 2 **a–d** Effects of sinusoidal electrical stimulation on action potentials in excitatory neurons. **a** HPW; **b** FR; **c** AP-Peak; **d** AMP

3.2 The Effects of Alternating Current Stimulation on the Firing Patterns of Inhibitory Neurons

Figure 3 illustrates the changes in membrane potential and firing frequency over time for inhibitory neurons under electrical stimulation intensities of 2, 4, 6, 8, 10, 15, 20, 25, 30, 35, 40, 45 pA. Under sinusoidal stimulation, the firing frequency of all three types of inhibitory neurons increased with higher stimulation intensity. Specifically:

FS (Fast Spiking) neurons exhibited the lowest threshold, responding at 4–8 pA. Their frequency increased with stimulation while maintaining stable waveform amplitude.

LTS (Low-Threshold Spiking) neurons showed weak initial responses but demonstrated synchronized increases in both frequency and amplitude with stronger stimuli. Beyond 20 pA, their frequency accelerated significantly with prominent sharp peaks.

LS (Late Spiking) neurons displayed low frequency and small amplitude at 4–8 pA. Their frequency increased steadily with intensity and rose substantially in both frequency and amplitude after 25 pA.

Figure 4 illustrates the dynamic responses of three types of inhibitory neurons (FS, LTS, LS) under sinusoidal current stimulation based on four parameters: HPW, FR, AP-Peak, and AMP.

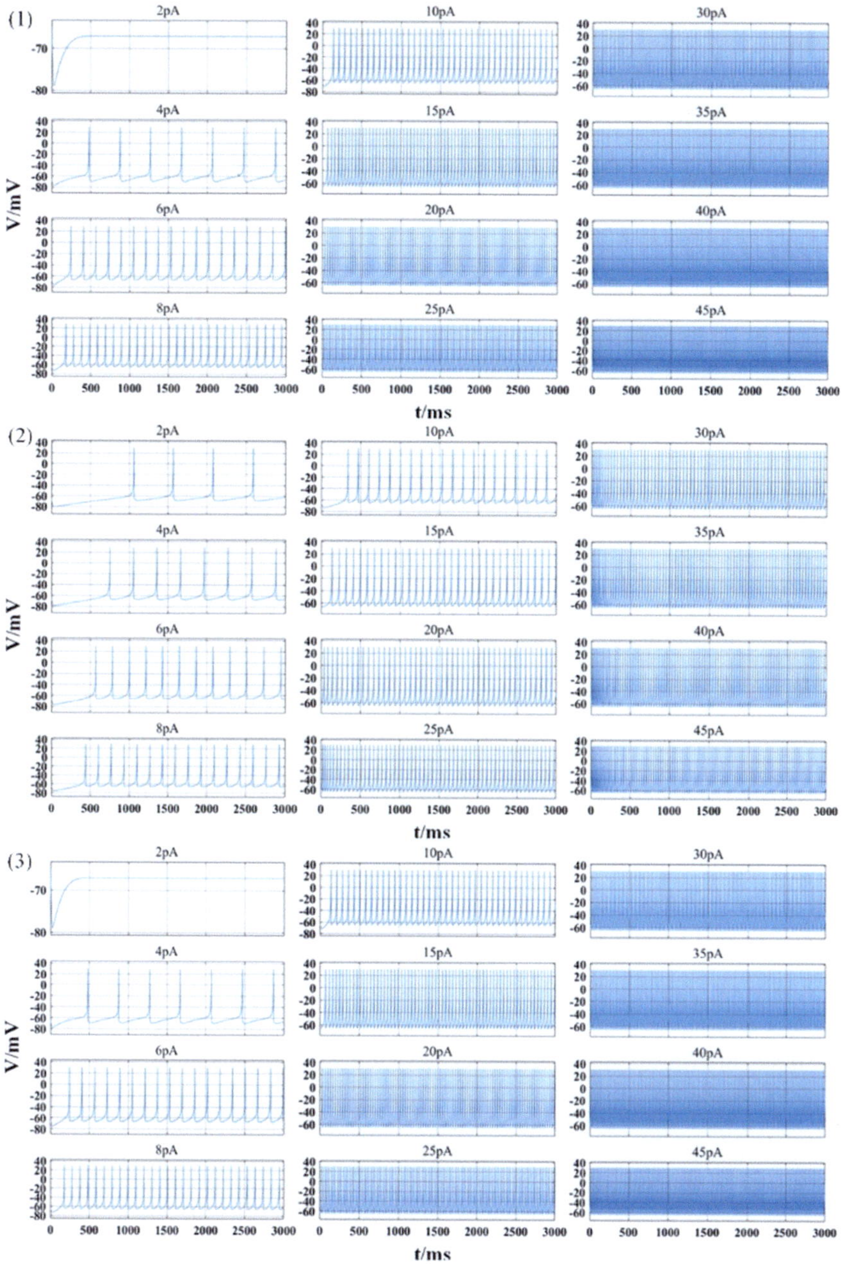

Fig. 3 Simulation results of action potentials in inhibitory neurons under sinusoidal electrical stimulation

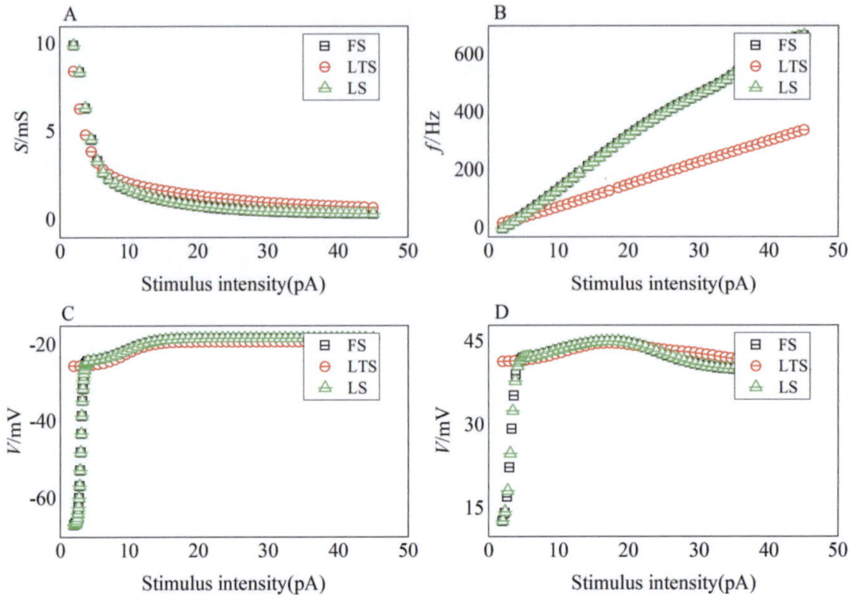

Fig. 4 **a–d** Effects of sinusoidal electrical stimulation on action potentials in Inhibitory neurons. **a** HPW; **b** FR; **c** AP-peak; **d** AMP

As shown in Fig. 4a, the half-peak width (HPW) of all three neuron types gradually decreased with increasing stimulation intensity and stabilized in the high-intensity range, indicating notable adaptive characteristics.

Figure 4b reveals that the firing rate (FR) of these neurons increased with stronger stimuli, though their response patterns differed: FS neurons exhibited an approximately linear and slow rise in frequency; LTS neurons showed rapid acceleration in the low-intensity range followed by gradual stabilization; LS neurons maintained a relatively lower firing rate with slower growth.

Figure 4c indicates that the action potential peak (AP-Peak) remained largely stable across varying stimulation intensities, with minimal fluctuations.

Figure 4d displays a nonlinear trend in action potential amplitude (AMP), which first increased and then decreased with higher stimulation intensity, where the reduction under high-intensity conditions may result from channel saturation or adaptive mechanisms.

Overall, under sinusoidal electrical stimulation, the three types of inhibitory neurons exhibited certain common response characteristics and trends in the four metrics—HPW, FR, AP-Peak, and AMP. First, the HPW generally stabilized under high stimulation intensities, reflecting the neurons' adaptive capacity to sustained strong stimuli. Second, the action potential peaks remained relatively consistent across all three neuron types throughout the stimulation intensity range. Finally, the membrane potential amplitude increased with stronger stimulation in all neurons, though the degree of adaptation varied among different neuronal types.

4 Conclusion

Under AC stimulation with intensities ranging from 2 to 45 pA, both excitatory and inhibitory neurons exhibited intensity-dependent responses. For excitatory neurons, the firing frequency increased with higher stimulation intensity, accompanied by a continuous shortening of the half-peak width (HPW). The action potential peak (AP-Peak) significantly increased once the stimulation exceeded a certain threshold, while the amplitude (AMP) initially rose and then markedly decreased.

Inhibitory neurons demonstrated a higher activation threshold in response to electrical stimulation. However, once the stimulation exceeded this threshold, their firing frequency also increased, along with a reduction in HPW and a continuous growth in AP-Peak. The AMP of inhibitory neurons showed an initial increase followed by a gradual decline. These results indicate that under suprathreshold stimulation, the trends in HPW, firing frequency, and action potential characteristics of inhibitory neurons align with those of excitatory neurons.

Acknowledgements This research was supported by the National Natural Science Foundation of China (Grant No. 52207251), and the Hebei Natural Science Foundation (Grant No. F2025201070).

References

1. Isaacson JS, Scanziani M (2011) How inhibition shapes cortical activity. Neuron 72(2):231–243
2. Ferguson KA, Cardin JA (2020) Mechanisms underlying gain modulation in the cortex. Nat Rev Neurosci 21(2):80–92
3. Altomare D, Benussi A, Cantoni V, Premi E, Rivolta J, Cupidi C, Martorana A, Santarnecchi E, Padovani A, Koch G, Borroni B (2023) Home-based transcranial alternating current stimulation (tACS) in Alzheimer's disease: rationale and study design. Alzheimer's Res Ther 15(1)
4. Borghini G, Candini M, Filannino C, Hussain M, Walsh V, Romei V, Zokaei N, Cappelletti M (2018) Alpha oscillations are causally linked to inhibitory abilities in ageing. J Neurosci 38(18):4418–4429
5. Riecke L, Formisano E, Herrmann CS et al (2015) 4-Hz transcranial alternating current stimulation phase modulates hearing. Brain Stimul 8(4):777–783
6. Jaušovec N, Jaušovec K (2014) Increasing working memory capacity with theta transcranial alternating current stimulation (tACS). Biol Psychol 96:42–47
7. Müller NG, Vellage AK, Heinze HJ et al (2015) Entrainment of human alpha oscillations selectively enhances visual conjunction search. PLoS ONE 10(11):e0143533
8. Kleinert ML, Szymanski C, Müller V (2017) Frequency-unspecific effects of θ-tACS related to a visuospatial working memory task. Front Hum Neurosci 11
9. Reato D, Rahman A, Bikson M, Parra LC (2013) Effects of weak transcranial alternating current stimulation on brain activity—a review of known mechanisms from animal studies. Front Hum Neurosci 7
10. Booth SJ, Taylor JR, Brown LJE, Pobric G (2022) The effects of transcranial alternating current stimulation on memory performance in healthy adults: a systematic review. Cortex 147:112–139
11. Izhikevich EM (2004) Which model to use for cortical spiking neurons 15(5):1063–1070

Modulation of Dynamic Functional Connectivity by rTMS During Working Memory Performance in AD

Jiaojiao Gao, Miaomiao Guo, Tian Wang, Lin Liu, Hao Wu, Yong Ji, Pan Wang, and Guizhi Xu

Abstract This study investigated the effects of repetitive transcranial magnetic stimulation (rTMS) on theta and alpha band dynamics within brain networks in Alzheimer's disease (AD) patients. We analyzed cognitive task state data during rTMS, focusing specifically on changes across the encoding, maintenance, and retrieval phases of working memory (WM). Dynamic functional connectivity (dFC) matrices were used to analyze graph-theoretic topological properties of brain networks. Furthermore, we examined the correlation between alterations in the topological properties of these dynamic brain networks and changes in WM performance

J. Gao · M. Guo (✉) · T. Wang · L. Liu · G. Xu (✉)
State Key Laboratory of Intelligent Power Distribution Equipment and System, School of Health Sciences and Biomedical Engineering, Hebei University of Technology, Tianjin, China
e-mail: gmm@hebut.edu.cn

G. Xu
e-mail: gzxu@hebut.edu.cn

Hebei Key Laboratory of Bioelectromagnetics and Neuroengineering, Hebei University of Technology, Tianjin, China

J. Gao
e-mail: 13292717628@163.com

T. Wang
e-mail: 18020047001@163.com

L. Liu
e-mail: 1092464221@qq.com

H. Wu · Y. Ji · P. Wang
Department of Neurology, Tianjin Huanhu Hospital, Tianjin, China
e-mail: xiaowu2pm@163.com

Y. Ji
e-mail: jiyongusa@126.com

P. Wang
e-mail: wpaofeier@163.com

Tianjin Key Laboratory of Cerebral Vascular and Neurodegenerative Diseases, Tianjin Dementia Institute, Tianjin, China

G. Mu et al. (eds.), *Proceedings of the 2nd Electrical Artificial Intelligence Conference, Volume 1*, Lecture Notes in Electrical Engineering 1587,
https://doi.org/10.1007/978-981-95-7932-7_10

metrics as well as clinical scale scores. Based on our results, we provide a comprehensive discussion on how rTMS influences the dynamic characteristics of WM brain networks in AD patients. This work offers a theoretical foundation for understanding the mechanisms by which rTMS may improve WM performance in AD, highlighting its potential therapeutic implications.

Keywords Repetitive transcranial magnetic stimulation · Alzheimer's disease · Electroencephalogram · Dynamic functional connectivity · Working memory

1 Introduction

Alzheimer's disease (AD) is a neurodegenerative disorder characterized by progressive cognitive decline, functional impairment, and neuropsychiatric symptoms [1]. Its pathology involves β-amyloid plaques, neurofibrillary tangles, and synaptic alterations [2]. Working memory (WM), a core deficit in AD, relies on dynamic network reorganization and is closely linked to functional connectivity disruption, an early AD manifestation [3–5]. With rising global aging trends, AD has become a major public health challenge [6, 7]. Current pharmacological treatments show limited efficacy and risk tolerance with prolonged use, underscoring the need for safe and effective alternatives or adjuncts [8]. Repetitive transcranial magnetic stimulation (rTMS), a non-invasive neuromodulation technique, has shown potential for improving cognitive symptoms in AD, though its mechanisms remain unclear [9–12].

Electroencephalography (EEG) enables real-time monitoring of brain activity, and dynamic functional connectivity (dFC) captures transient network states with high temporal resolution [13, 14]. Theta (4–8 Hz) and alpha (8–13 Hz) oscillations are critically involved in WM processes: theta facilitates prefrontal-hippocampal interactions during encoding and retrieval, while alpha regulates attention and inhibits interference [15]. Theta desynchronization and pathological alpha hypersynchrony reflect early network disruption in AD [16, 17]. These frequency-specific abnormalities making these bands ideal for probing rTMS-induced network changes.

In this study, we combined dFC and graph theory to track network topology changes across WM stages in AD patients receiving rTMS. We hypothesized that rTMS drives dynamic network reorganization, and examined how these changes correlate with cognitive performance. This network-level approach offers new insights into rTMS mechanisms and supports therapy optimization.

Table 1 Summary of the patients' clinical characteristic data (* < 0.05, ** < 0.01)

	Before-rTMS	After-rTMS	t/χ^2-value	p-value
Gender(M:F)	8/12			
Age/years of disease/education	67.26 ± 4.39/6 ± 1.42/10.05 ± 2.89			
MMSE	19.20 ± 6.31	21.25 ± 5.36	−3.466	0.003**
MoCA	15.15 ± 6.69	18.95 ± 6.82	−6.278	0.002**
RAVLT-learn	19.70 ± 6.94	22.80 ± 10.99	−1.785	0.090
RAVLT-immediate recall	5.85 ± 3.79	7.50 ± 3.58	−3.116	0.006**
DST	11.10 ± 1.80	11.05 ± 1.82	0.006	0.909
BNT	9.70 ± 4.11	11.85 ± 4.93	−3.929	0.002**
NPI	6.25 ± 6.96	3.40 ± 3.68	1.603	0.125
ADL	26.62 ± 8.55	25.00 ± 6.78	1.400	0.178
HAMD	4.55 ± 3.59	2.60 ± 2.56	2.534	0.020*

2 Material and Methods

2.1 Participants and Neuropsychological Assessment

This study enrolled 20 AD patients from Tianjin Huanhu Hospital (China). The experimental procedures, including participant inclusion and ethical review, were conducted in accordance with those detailed in the prior published study [18]. This study protocol was approved by the Institutional Review Board of Hebei University of Technology, Tianjin, China (No. HEBUThMEC2024009) and conducted in accordance with the Declaration of Helsinki.

Cognitive assessments were administered before and after treatment, including the Mini-Mental State Examination (MMSE), Montreal Cognitive Assessment (MoCA), Rey Auditory Verbal Learning Test (RAVLT), Digit Span Test (DST), Boston Naming Test (BNT), Neuropsychiatric Inventory (NPI), Activity of Daily Living Scale (ADL) and Hamilton Depression Rating Scale (HAMD). The subjects demographics and clinical cognitive assessments results were presented in Table 1.

2.2 rTMS Protocol

The rTMS protocol was performed using a Magstim Rapid2 stimulator (Magstim, UK) with a figure-of-eight coil, and the stimulation target area is the left dorsolateral prefrontal cortex. The coil was positioned tangentially to the sagittal plane at 45°. Patients underwent 14 daily sessions, each comprising 1600 pulses (40 trains of 2 s

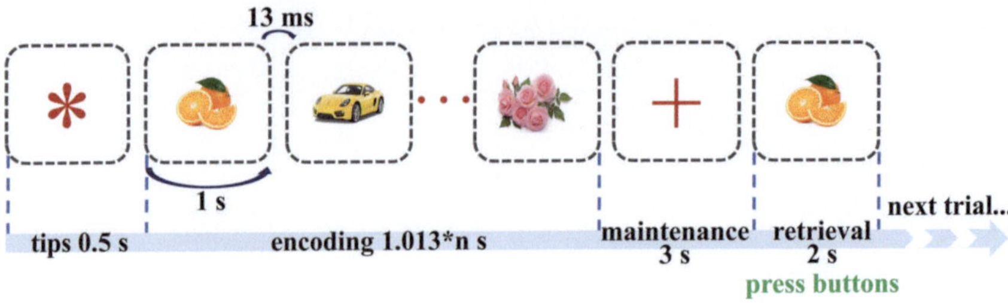

Fig. 1 Task flow of WM task paradigm

duration, 28 s intervals) delivered at 90% resting motor threshold. Treatments were conducted in a controlled quiet environment.

2.3 Working Memory Task Paradigm

WM was evaluated using an n-back task (Fig. 1). Each trial consisted of encoding (sequential image presentation), maintenance (active memorization), and retrieval (matching probe to memorized images) phases. The task included four load conditions (1- to 4-back), with recorded measures of response time (probe onset to correct key press) and accuracy (percentage correct).

2.4 EEG Data Acquisition and Preprocessing

EEG was recorded in a quiet environment using a 64-channel wireless system (Neuracle, China) with electrodes positioned following the 10–20 system. Impedance was kept below 10 kΩ and sampling rate was 1000 Hz. Signals were acquired at four timepoints: before rTMS (baseline), after the first (immediate), seventh (stage), and fourteenth (final) sessions. Preprocessing was conducted in EEGLAB [19], involving bandpass (0.5–45 Hz) and notch (48.5–51.5 Hz) filtering, artifact removal through independent component analysis with manual inspection, followed by re-referencing and final filtering.

2.5 Dynamic Networks Analysis

This study constructed dFC networks using a sliding window approach with a 2000 ms window and 50 ms step size to optimize temporal resolution and stationarity. Preprocessed EEG data were segmented using 50% overlapping windows to ensure temporal continuity.

$$time_l = 1 + M \times (N - 1) \tag{1}$$

$$time_r = Z + M \times (N - 1) \tag{2}$$

where Z is the window length, M is the step size, $time_l$ denotes the left boundary index, and $time_r$ denotes the right boundary index.

Phase locking value (PLV) analysis was then applied to each segment to generate symmetric functional connectivity matrices. To mitigate inter-subject variability and reduce dimensionality, principal component analysis was performed on concatenated individual matrices, from which eigenvectors and eigenvalues were derived.

$$\mathrm{M} = [\mathrm{A1}, \mathrm{A2}, \mathrm{A3} \ldots, \mathrm{A20}] \tag{3}$$

$$\mathrm{MM}^{\mathrm{T}}\mathrm{W} = \lambda \mathrm{W} \tag{4}$$

where λ is a diagonal matrix of eigenvalues and W is an orthogonal matrix whose columns are the corresponding eigenvectors.

The top k eigenvectors corresponding to the largest eigenvalues (each of dimensio: $(\mathrm{N}^2 - \mathrm{N})/2 \times 1$) were selected and symmetrically reconstructed to derive the $\mathrm{N} \times \mathrm{N}$ dominant connectivity matrices, as illustrated in Fig. 2.

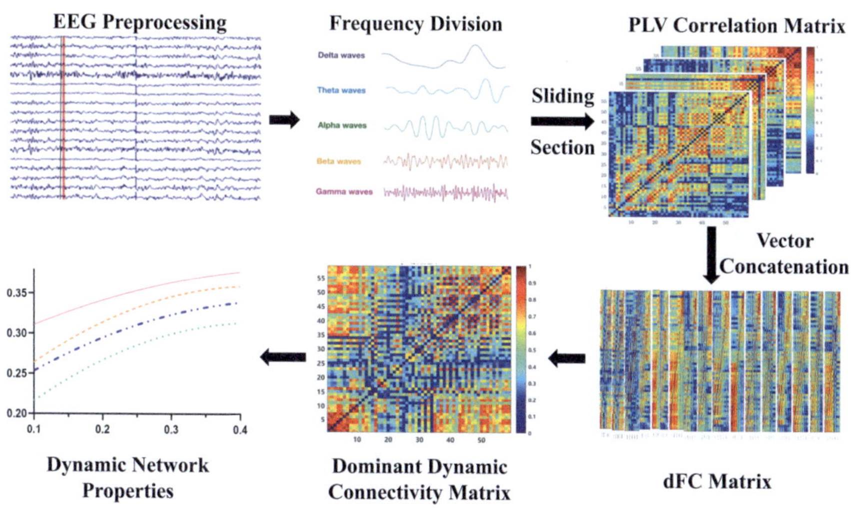

Fig. 2 Flow chart of functional brain network construction

2.6 Statistical Analysis

Statistical analyses were performed using SPSS 27.0 (IBM, USA) and GraphPad Prism 10.1.2. Data are expressed as mean ± standard deviation or median with interquartile range according to distribution. Group comparisons utilized t-tests or chi-square tests, while Spearman correlation assessed relationships between network topology alterations, WM behavior, and clinical scores. A significance threshold of $p < 0.05$ was applied.

3 Results

3.1 Participants Characteristics

The demographic information of the participants is presented in Table 1. Post-rTMS assessment demonstrated significant improvements in cognitive functions alongside a reduction in depressive symptoms.

3.2 Results of Dynamic Network Connectivity Analysis

dFC analysis during rTMS revealed progressive network reorganization across WM phases (see Fig. 3). AD patients showed initial within-region FC enhancement that evolved into broader integration, notably in fronto-occipital pathways. While baseline FC was higher during encoding, rTMS improved FC across all phases, especially maintenance and retrieval. Encoding and maintenance primarily enhanced within-region connectivity, whereas retrieval strengthened between-region links.

For robustness, we analyzed network properties under varying thresholds (see Fig. 4). Higher thresholds increased the clustering coefficient and shortened path length. Theta band clustering decreased during encoding but increased in later phases, while alpha band clustering rose throughout. Path length shortened consistently across both bands and all phases. During final-stage WM retrieval, the alpha band showed the greatest clustering increase and shortest path length.

3.3 Results of rTMS Effects on Network Properties

Network topology analysis using a 20% dFC threshold revealed frequency- and stage-specific rTMS effects (see Fig. 5). During encoding, theta band exhibited progressively reduced clustering and shortened path length, while alpha band showed decreased clustering with prolonged path length. In maintenance, theta clustering

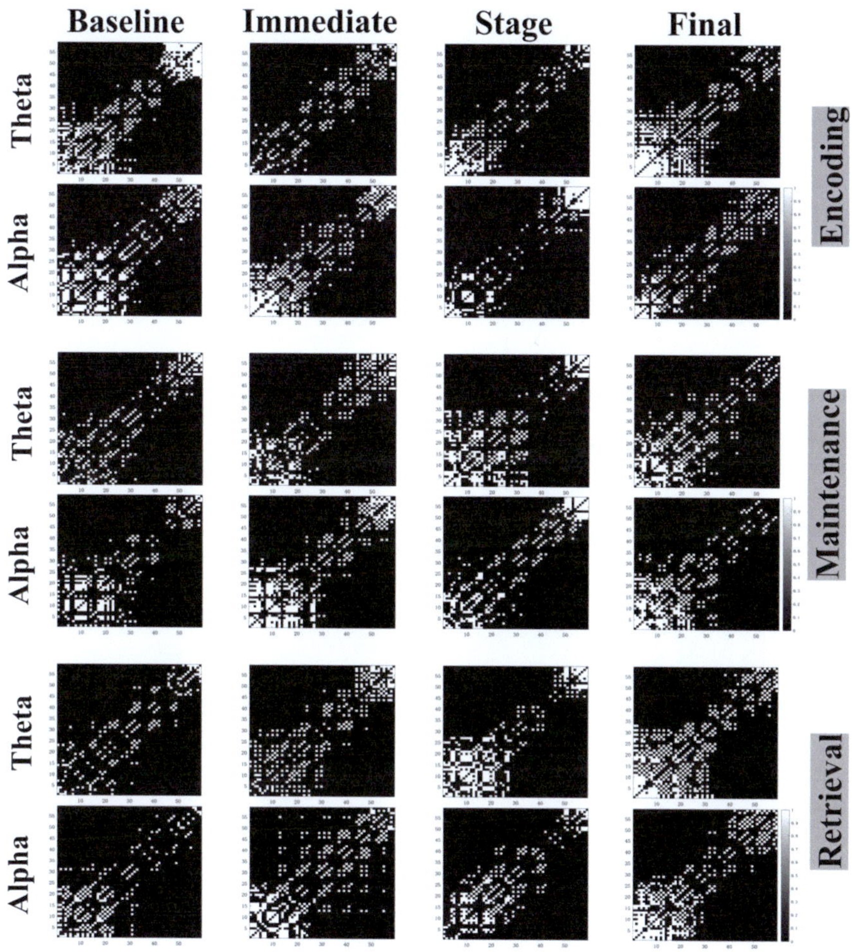

Fig. 3 Dynamic binary connectivity matrices during WM tasks in theta and alpha bands. Axes represent node numbers

increased immediately while path length shortened only finally, alpha path length progressively decreased throughout. During retrieval, theta demonstrated immediate path reduction and staged clustering increase, while alpha showed sustained clustering enhancement and path shortening.

To further investigate rTMS regulation of executive function in specific brain regions of AD patients, we analyzed dynamic changes in local FC properties across WM stages. Regional FC analysis revealed distinct patterns: encoding involved immediate frontal-parietal degree increases and widespread stage effects, maintenance featured frontal enhancements, while retrieval showed progressive involvement from frontal to posterior regions across both bands (see Fig. 6).

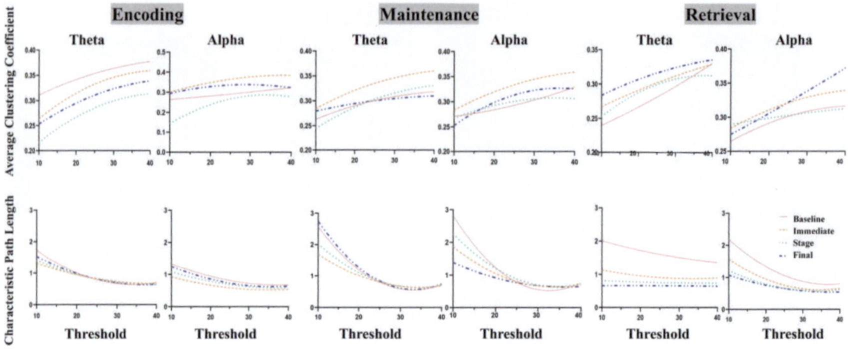

Fig. 4 Comparison of dynamic network properties in the principal connectivity matrices during WM tasks. Subject baseline (red solid line), immediate (orange dashed), stage (blue dashed), and final (green dashed) effects are shown

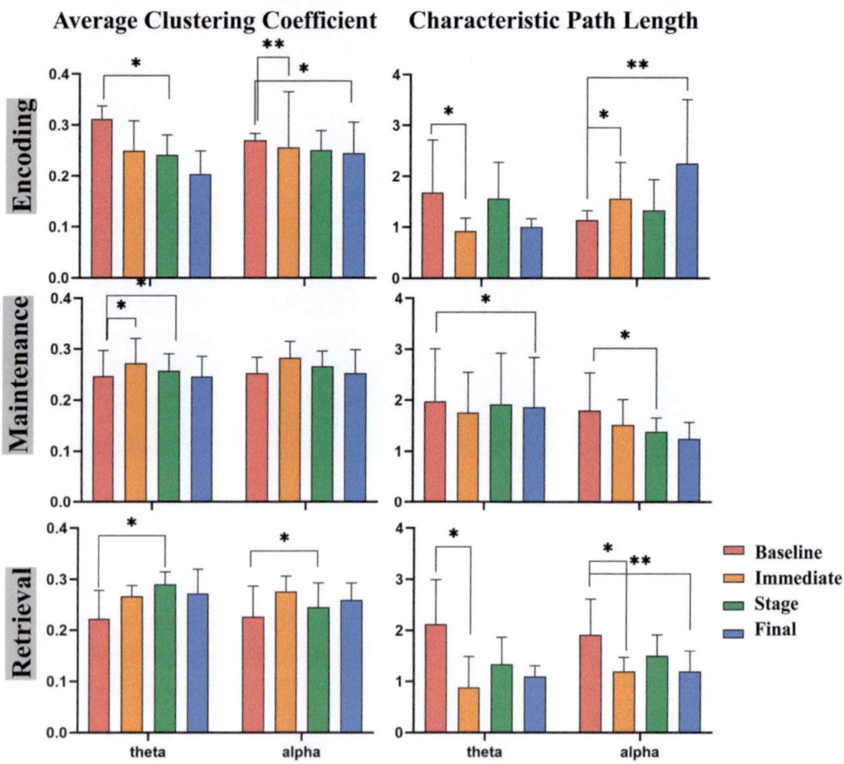

Fig. 5 Dynamic global network properties of the main connectivity matrix during WM

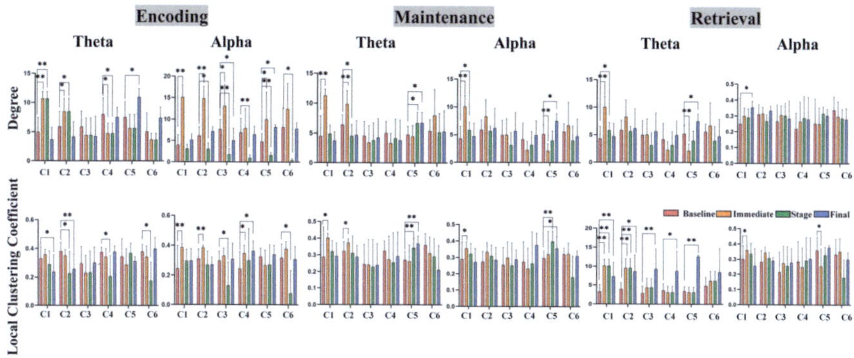

Fig. 6 Dynamic local network properties of the main connectivity matrix during WM. C1–C2: frontal lobe; C3: central area; C4: parietal lobe; C5: occipital lobe; C6: temporal lobe

3.4 Results of Correlation Analysis

We analyzed correlations between network properties and behavioral/clinical measures. Across theta and alpha bands, WM accuracy increased with higher node degree and clustering coefficient, and decreased with longer characteristic path length. WM judgment time showed opposite trends (see Fig. 7).

Correlation analysis using pre-to-post rTMS change values revealed significant associations during WM encoding. Theta clustering and path length negatively correlated with MMSE (r = −0.930), alpha clustering negatively correlated with MMSE (r = −0.950) and positively with ADL (r = 0.560). During maintenance, alpha clustering positively correlated with MMSE (r = 0.907) and DST (r = 0.625), while path length negatively correlated with MoCA (r = −0.576) and RAVLT-Immediate Recall

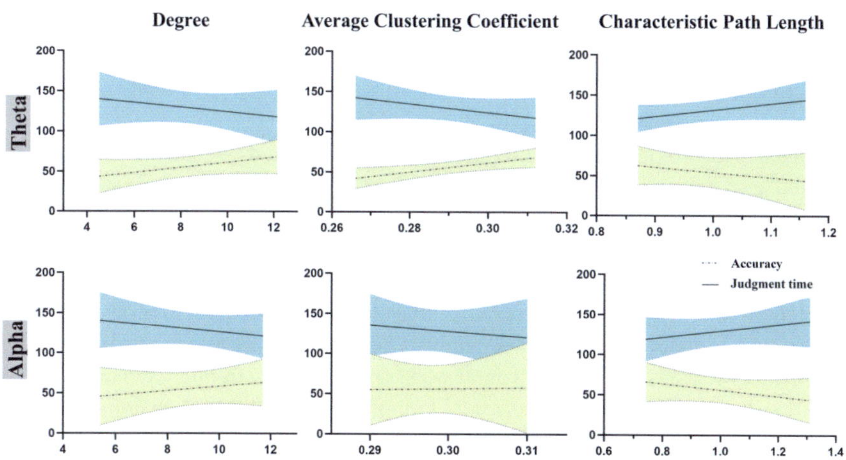

Fig. 7 Correlation analysis between network properties and WM task performance

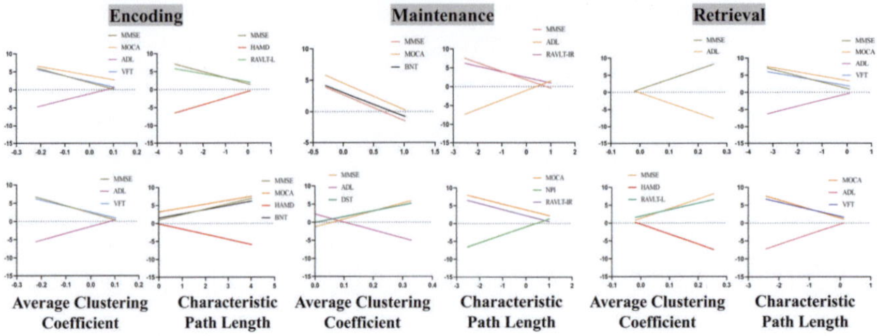

Fig. 8 Correlation analysis between network properties in theta (top) and alpha (bottom) bands with clinical assessment scores

(r = −0.639). During retrieval, theta clustering positively correlated with MMSE (r = 0.854), while alpha path length negatively correlated with MoCA (r = −0.890) (all $p < 0.05$, see Fig. 8).

4 Discussion

Our study investigated the dynamic effects of rTMS on AD patients using WM task-related EEG and dFC analysis across four treatment stages. We characterized rTMS-induced alterations in functional network properties and their association with cognitive performance.

Consistent with previous findings on rTMS efficacy, we observed significant improvements in general cognition (MMSE, MoCA) and depressive symptoms (HAMD), alongside specific gains in verbal memory (RAVLT) and language (BNT). These clinical outcomes provide context for the observed dFC changes. dFC analysis revealed progressive network integration enhancement. Baseline frontal hyperconnectivity during encoding implied compensatory mechanisms [20], while cumulative improvements reflected neuroplastic changes [21].

Focusing on frequency-specific effects [22], theta-band analysis showed rTMS reduced excessive local clustering during encoding while enhancing it during retrieval, potentially rectifying AD's pathological network topology [23]. Alpha-band networks exhibited uniformly increased clustering and reduced path length across WM stages, peaking during retrieval. The robust inverse correlation between path length and WM accuracy confirms shorter global pathways facilitate efficient information transfer [23]. Network analysis revealed rTMS selectively modulated WM-critical regions. Increased frontal node degree across frequency bands during encoding/retrieval aligns with prefrontal targeting [24]. Enhanced occipital connectivity suggests improved visuospatial processing, while parietal theta clustering may optimize attentional allocation [20].

Significant correlations between network metrics and clinical scales validated their functional relevance (Fig. 8). Negative theta clustering-MMSE correlation during encoding indicated excessive local connectivity impedes cognition, while positive alpha clustering-MMSE correlation during retrieval suggested alpha synchronization supports memory consolidation [25]. Crucially, shortened path length correlated with improved MoCA and RAVLT scores, affirming that network efficiency gains translate to cognitive benefits [20, 26].

Study limitations include the relatively small sample size, warranting future validation in larger cohorts. Additionally, while EEG provided valuable dynamic connectivity data, multimodal neuroimaging integration would offer complementary structural insights into rTMS mechanisms in AD.

5 Conclusion

rTMS ameliorated AD-related network defects by dynamically reconfiguring dFC: enhancing global integration (reduced characteristic path length), optimizing local segregation (variable clustering), and amplifying alpha-band synchronization. These network-level changes correlated with improved cognition and affect in AD patients, positioning dFC as a biomarker for evaluating the efficacy of rTMS.

Acknowledgements This study was funded by the National Key R&D Program of China (2022YFC2402200), the Project supported by the Natural Science Foundation of Hebei Province, China (F2024202085), the Funds for International Cooperation and Exchange of the National Natural Science Foundation of China (52320105008), Tianjin Health Science and Technology Program of China (TJWJ2022MS032), and Tianjin Health Research Project (TJWJ2024QN062), Tianjin Science and Technology Major Project (24ZXGQSY00070), Tianjin Municipal Education Commission Research Project (2024ZD060), and Tianjin Natural Science Foundation (25JCLMJC00070).

References

1. Scheltens P, De Strooper B, Kivipelto M et al (2021) Alzheimer's disease. Lancet 397(10284):1577–1590
2. Kuznetsov A (2025) Evaluating the combined neurotoxicity of amyloid beta and tau oligomers in Alzheimer's disease: a novel cellular-level criterion. J Biomech Eng 22:1–46
3. Thomas B, Christophel P, Christiaan K et al (2016) The distributed nature of working memory. Trends Cogn Sci 21(2):111–124
4. Zhao C, Huang W, Feng F et al (2022) Abnormal characterization of dynamic functional connectivity in Alzheimer's disease. Neural Regen Res 17(9):2014–2021
5. Stanley ML, Simpson SL, Dagenbach D et al (2015) Changes in brain network efficiency and working memory performance in aging. PLoS ONE 10(4):e0123950
6. Xu Y, Wang J, Wang H et al (2023) Data and prevention-control strategies of Alzheimer's disease in China. J Alzheimer's Dis Relat Disord 6(3):175–192+173 (in Chinese)
7. Rostagno AA (2022) Alzheimer's disease pathogenesis. Int J Mol Sci 24(1):107

8. Szeto JY, Lewis SJ (2016) Current treatment options for Alzheimer's disease and Parkinson's disease dementia. Curr Neuropharmacol 14(4):326–338
9. Barker AT, Jalinous R, Freeston IL (1985) Freeston, Non-invasive magnetic stimulation of human motor cortex. Lancet 325(8437):106–1107
10. Siebner HR, Funke K, Aberra AS et al (2022) Transcranial magnetic stimulation of the brain: what is stimulated—a consensus and critical position paper. Clin Neurophysiol 140:59–97
11. Wang X, Mao Z, Ling Z et al (2020) Repetitive transcranial magnetic stimulation for cognitive impairment in Alzheimer's disease: a meta-analysis of randomized controlled trials. J Neurol 267(3):791–801
12. Shang YQ, Xie J, Peng W et al (2018) Network-wise cerebral blood flow redistribution after 20 Hz rTMS on left dorso-lateral prefrontal cortex. Eur J Radiol 101:144–148
13. Gavaret M, Iftimovici A, Pruvost-Robieux E et al (2023) EEG: current relevance and promising quantitative analyses. Revue Neurol 179(4):352–360
14. Fu Z, Caprihan A, Chen J et al (2019) Altered static and dynamic functional network connectivity in Alzheimer's disease and subcortical ischemic vascular disease: shared and specific brain connectivity abnormalities. Hum Brain Mapp 11:3203–3221
15. Hsieh LT, Ranganath C (2014) Frontal midline theta oscillations during working memory maintenance and episodic encoding and retrieval. Neuroimage 85:721–729
16. Jensen O et al (2002) Oscillations in the alpha band (9–12 Hz) increase with memory load during retention in a short-term memory task. Cereb Cortex 12(8):877–882
17. Moretti DV (2015) Theta and alpha EEG frequency interplay in subjects with mild cognitive impairment: evidence from EEG, MRI, and SPECT brain modifications. Front Aging Neurosci 7:31
18. Wang T, Guo M, Wu H et al (2025) rTMS modulates early AD progression via synergistic brain network reorganization and peripheral biomarker dynamics. Geroscience
19. Delorme A, Makeig S (2004) EEGLAB: an open source toolbox for analysis of single-trial EEG dynamics including independent component analysis. J Neurosci Methods 134(1):9–21
20. Zhou B, Yao H, Wang P, et al (2015) Aberrant functional connectivity architecture in Alzheimer's disease and mild cognitive impairment: a whole-brain, data-driven analysis. Biomed Res Int 495375
21. Schneider D, Göddertz A, Haase H et al (2019) Hemispheric asymmetries in EEG alpha oscillations indicate active inhibition during attentional orienting within working memory. Behav Brain Res 359:38–46
22. Alekseichuk I et al (2016) Spatial working memory in humans depends on theta and high gamma synchronization in the prefrontal cortex. Curr Biol 26(12):1513–1521
23. Dai ZJ, Lin QX, Li T et al (2019) Disrupted structural and functional brain networks in Alzheimer's disease. Neurobiol Aging 75:71–82
24. Matias MG, Maciel DG, França IM et al (2022) Transcranial direct current stimulation associated with functional exercise program for treating fibromyalgia: a randomized controlled trial. Arch Phys Med Rehabil 103(2):245–254
25. Paldino MJ, Zhang W, Chu ZD et al (2017) Metrics of brain network architecture capture the impact of disease in children with epilepsy. Neuroimage Clin 13:201–208
26. Wang J, Huang L, Wei L et al (2021) Factors affecting the efficacy of repetitive transcranial magnetic stimulation for patients with Alzheimer's disease. J Zhejiang Univ Med Sci 50(3):383–389 (in Chinese)

Artificial Intelligence Accelerates Low Carbon Transformation of Thermal Power Plants

Yan Zhang, Yuetong Zhu, Zhentao Zhu, Xinyang Xia, Zi Yuan, and Jican Liu

Abstract Under the 'dual carbon' goals, coal-fired power plants face severe challenges in their low-carbon transition. Key decarbonisation pathways, such as flexible retrofitting, fuel blending, and carbon capture, utilisation, and storage (CCUS), are all hindered by complex processes, conflicting objectives, and high costs. This paper aims to systematically explain how artificial intelligence (AI) can provide data-driven solutions to overcome these challenges and accelerate the low-carbon transition of coal-fired power plants. This review employs bibliometric tools to systematically screen literature and analyse research hotspots and evolving trends. The study constructs a 'three-stage evolution model' for AI-enabled coal-fired power plants, revealing a paradigm shift from local optimisation to autonomous evolution. The study found that the core of AI lies in transforming industrial challenges that are difficult to model precisely using traditional methods into data problems that are learnable, predictable, and optimisable. In terms of flexible operation, AI achieves combustion optimisation and stable control through high-precision prediction models and advanced controllers. In fuel blending, AI establishes intelligent combustion models to optimise boiler efficiency, emissions, and fuel supply chain management. In the CCUS field, AI permeates the entire 'capture-transport-storage-utilisation' chain, fundamentally reshaping its technical and economic feasibility through energy consumption optimisation, new material discovery, and efficiency improvements. AI is driving the evolution of thermal power plants from local optimisation to a plant-wide, full-life-cycle intelligent operation paradigm based on digital twins.

Keywords AI · Thermal power plant · Low-carbon transition · CCUS · Ultra-supercritical · Machine learning · Coal-fired power plant

---

Y. Zhang · Z. Zhu
School of Business, Nanjing Institute of Technology, Nanjing, China

Y. Zhu (✉) · X. Xia · Z. Yuan · J. Liu
School of Power Engineering, Nanjing Institute of Technology, Nanjing, China
e-mail: y00450240650@njit.edu.cn

G. Mu et al. (eds.), *Proceedings of the 2nd Electrical Artificial Intelligence Conference, Volume 1*, Lecture Notes in Electrical Engineering 1587,
https://doi.org/10.1007/978-981-95-7932-7_11

1 Introduction

The severe challenge of global climate change has become one of the most urgent issues facing human society [1]. In response, nations worldwide have set "dual carbon" goals, aiming for net-zero emissions. Thermal power, particularly coal-fired power, remains the "ballast stone" for energy security. However, it is also the world's largest single source of carbon emissions [2]. As renewable energy grows, coal plants must shift from "baseload" to "regulatory" power sources, a role for which they were not designed. This transition presents three core dilemmas:

1. **Flexibility Dilemma**: Deep peaking and rapid load changes cause deteriorating combustion stability, efficiency decay, and increased equipment safety risks, creating a "safety-economy" conflict.
2. **Co-firing Dilemma**: Co-firing biomass or ammonia introduces fuels with vastly different properties, leading to combustion instability, pollutant trade-offs (e.g., NOx paradox), and material corrosion, creating a "clean-efficient" balancing problem.
3. **CCUS Dilemma**: CCUS, a key deep decarbonization technology, is trapped in a "vicious cycle" of high capital costs, massive energy penalties, and substandard operational performance, resulting in an "energy-cost" deadlock.

Traditional methods have reached their limits in solving these complex, multi-objective problems. Artificial intelligence (AI) offers a new, data-driven path. This paper systematically reviews how AI can break these three dilemmas, proposing a "three-stage evolution model" for AI empowerment, shifting the paradigm from local optimization to autonomous evolution [3].

2 Review Methodology and Research Landscape

2.1 Systematic Literature Screening

This study utilizes the indexes in the Web of Science database as the main source of statistical data. A specific search was conducted using TS = ((("AI" OR "Artificial Intelligence" OR "Machine Learning" OR "Deep Learning" OR "neural network" OR "digital twin") AND (("Ultra-Supercritical (USC)" OR "Double-Reheat USC" OR "Advanced Supercritical") OR ("Ammonia-Mixed Combustion" OR "Ammonia Co-firing" OR "Ammonia Combustion" OR "Ammonia-Mixed Combustion") OR ("Biomass Co-firing" OR "Biomass Blending" OR "Biomass Co-combustion" OR "Biomass Gasification Co-firing") OR ("CCUS" OR "Carbon Capture, Utilization and Storage" OR "Geological Sequestration" OR "EOR" OR "Mineral Carbonation" OR "Bioenergy with CCUS" OR "Negative Emission Technology (NET)" OR "Carbon Management Technology" OR "oxygen-rich combustion"))) field. To ensure the timeliness and relevance of the research, the scope of this literature review is limited to documents published between January 1, 2000, and July 1, 2025. We particularly focus on the latest research progress since 2016.

In the initial search, 580 relevant documents were retrieved. After an initial screening of titles and abstracts, 113 documents with low relevance to the research topic were excluded, resulting in 467 documents. Subsequently, by reading the abstracts of these 467 documents, we selected 37 highly relevant core documents as the theoretical basis and argumentation support for the subsequent chapters (Sects. 3–7).

2.2 Bibliometric-Based Analysis of Research Hotspots and Evolutionary Trends

To more objectively gain insight into the full picture and dynamics of research in the field of AI-enabled low-carbon transformation of thermal power plants, this paper further employs bibliometric tools to analyze the selected literature.

Figure 1 clearly shows the annual and cumulative publication growth trends for research related to AI-enabled low-carbon transformation of coal-fired power plants from 2016 to 2025. As shown, research activity in this field shows a significant growth trend, especially after 2019, with an accelerated increase in publications. Specifically, the annual publications grew steadily from about 10 in 2016 to a peak of 135 in 2024, and although only the first half of 2025 was counted, it remained at a high level of 100. At the same time, the cumulative publications show an even steeper upward curve, rapidly accumulating from just 10 in 2016 to 580 in 2025. This significant growth trend strongly reflects the academic community's growing attention to and recognition of artificial intelligence in solving the dilemmas of low-carbon transformation in coal-fired power plants. It also indicates its potential to become a hot frontier in future energy research.

2.3 Keyword Co-Occurrence Network and Research Theme Clustering

To further identify the core research hotspots and their interrelationships in this field, this paper conducted a co-occurrence network analysis of the keywords from the included literature, with the results shown in Fig. 2. In the figure, different colored clusters represent different research theme groups, node size reflects the frequency of keyword occurrence, and the lines between nodes indicate the co-occurrence strength.

In summary, these keyword clusters are highly consistent with the three major low-carbon transition pathways proposed in the introduction: flexible retrofitting of thermal power units, co-firing of multiple fuels, and carbon capture, utilization, and storage (CCUS). This not only validates the scientific and systematic nature

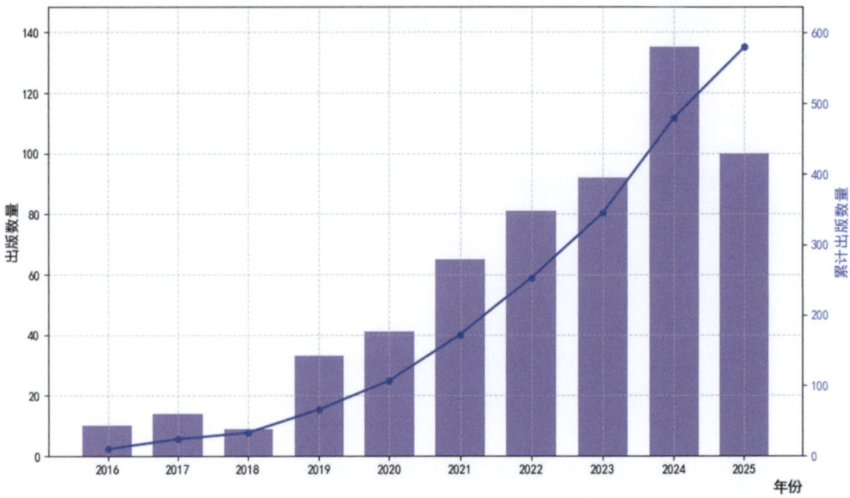

Fig. 1 Annual publications and cumulative trends in AI in the field of low-carbon transformation of thermal power generation

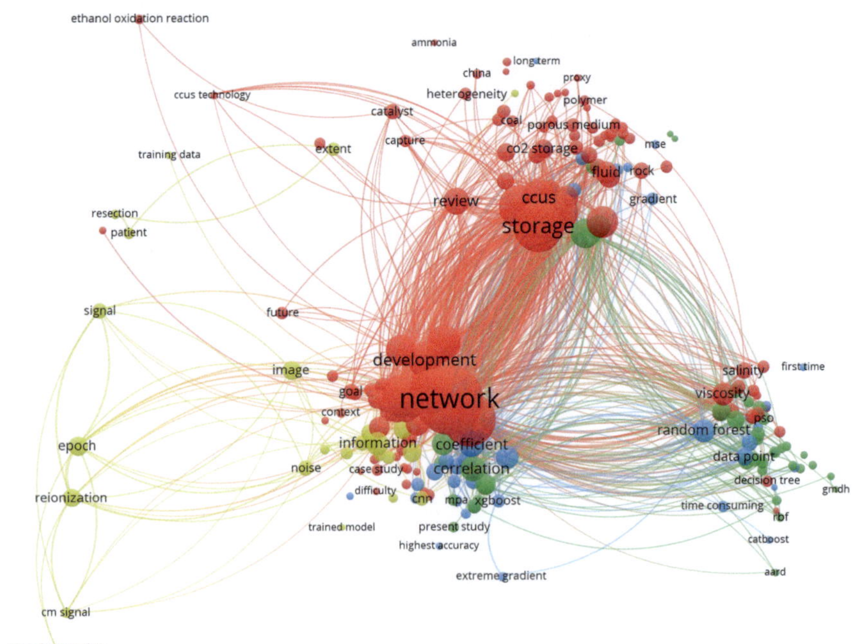

Fig. 2 Co-occurrence network diagram of hot keywords in research on AI-enabled low-carbon transformation of coal-fired power plants

of this review's framework but also further emphasizes that artificial intelligence is penetrating every core segment of the low-carbon transformation in thermal power plants from a technical level, becoming a key force in breaking traditional bottlenecks and promoting industry paradigm shifts.

3 AI-Enabled Flexibility Retrofitting and Cracking the "Safety-Economy" Conflict

3.1 *The Dilemma of Deep Peaking*

Ultra-supercritical (USC) units, designed for stable baseload operation, are now forced to act as "regulators" [4], leading to a severe "safety-economy" conflict. This is characterized by:

(1) **Increased equipment safety risks** due to thermal stress and fatigue from frequent load changes [5];
(2) **Deteriorating operational stability and efficiency** as the complex, non-linear combustion process becomes difficult to control outside its design point [6];
(3) **Increased difficulty in pollutant control** under variable conditions.

3.2 *Predictive Control: From Passive Response to Proactive Risk Avoidance*

AI's core advantage is its ability to build high-precision predictive models from complex data, shifting control from passive to proactive.

Combustion Optimization and Efficiency: AI algorithms like Artificial Neural Networks (ANN) are used to learn historical data (DCS), creating a precise map between control parameters (e.g., air–fuel ratio) and key performance indicators (KPIs) (e.g., efficiency, NOx). The mapping is a non-linear function:

$$y = \phi\left(\sum_{i=1}^{n} w_i x_i + b\right) \tag{1}$$

Based on this, evolutionary algorithms (e.g., Genetic Algorithms) find the optimal control settings by minimizing a multi-objective function J:

$$\max J(X) = w_1 \cdot f_\eta(X) - w_2 \cdot f_{NO_x}(X) - w_3 \cdot f_{CO}(X) \tag{2}$$

For example, research by Ashraf W M et al. demonstrated that AI-based parameter tuning could improve thermal efficiency by 3.12 percentage points [7]. More advanced online optimization systems have achieved simultaneous improvements in efficiency, emissions, and safety.

Robust Control and Dynamic Response: AI is used to model and control the complex, multi-variable Coordinated Control System (CCS). Hou G et al. proposed a novel fuzzy neural network (FNN) for 1000 MW unit CCS modeling, achieving satisfactory accuracy [6]. This allows for more precise and robust control over a wide range of operating loads.

Equipment Health Management: AI-driven predictive maintenance (PdM) is critical for mitigating safety risks. For instance, Yin L and Zhou H et al. developed a modal decomposition integrated model (MDIM) using deep learning (ResNet, Transformer) to provide high-precision, multi-step prediction of reheater tube temperatures, giving managers valuable time to prevent tube bursts [8]. This shifts maintenance from passive repair to active prevention, ensuring long-term safety and economy.

4 AI-Enabled Fuel Co-Firing and Mastering the "Clean-Efficient" Multi-Dimensional Balance

Co-firing low-carbon fuels like biomass and ammonia in existing boilers is a promising transitional strategy [9–11]. However, introducing fuels with vastly different properties creates a multi-dimensional "clean-efficient" dilemma [12].

4.1 Inherent Challenges of Co-Firing Technology

The primary challenges include:

1. **Combustion Instability**: Biomass's low heat value and ammonia's slow flame speed can destabilize the furnace, reduce efficiency [13–15], and even risk flameout [16, 17].
2. **Pollutant "Trade-off Dilemma"**: Ammonia co-firing, while carbon-free, can paradoxically increase NOx emissions ("carbon reduction, nitrogen increase") [18, 19]. Biomass co-firing alters PM and SOx emissions in complex ways [10].
3. **Material Corrosion**: Biomass is often rich in chlorine and alkali metals, causing severe high-temperature corrosion and s lagging [14].
4. **Fuel Variability and Economics**: The inconsistent quality of biomass and the high cost of green ammonia pose significant logistical and economic challenges.

4.2 AI in Co-Firing Full Value Chain Optimization

AI's non-linear modeling and global optimization capabilities are key to managing this multi-dimensional problem.

4.2.1 Intelligent Combustion Control and Multi-Objective Optimization

AI is used to build high-fidelity "surrogate models" (e.g., using ANN) that predict boiler KPIs (efficiency, NOx, CO) much faster than traditional CFD [16]. For example, Grądz Ż et al. used neural networks to classify flame states with over 99% accuracy, effectively monitoring combustion stability.

These models are then combined with multi-objective genetic algorithms (MOGA) to find the "Pareto optimal" balance between conflicting goals. The objective function:

$$J = \sum_{i=1}^{n} w_i \cdot f_i(x) \tag{3}$$

allows operators to adjust weights (w_i) based on priorities (e.g.,efficiency versus emissions).

4.2.2 Precise Pollutant Emission Prediction and Control

AI enables "proactive control" of emissions. The performance of these predictive models is often measured by the Coefficient of Determination (R^2):

$$R^2 = 1 - \frac{\sum_{i=1}^{n} (y_i - \hat{y}_i)^2}{\sum_{i=1}^{n} (y_i - \bar{y}_i)^2} \tag{4}$$

To tackle the ammonia-NOx paradox, high-precision models have been developed. A study by Liou J L et al. found a stacked generalization ensemble model (SGEM) achieved an R^2 as high as 0.944 for NOx prediction [20].

4.2.3 Intelligent Fuel Management and Asset Health

AI's application extends across the entire value chain:

Intelligent Fuel Blending: ANN models can determine the optimal blend of coal and biomass to balance calorific value, ash, and chlorine content [21]. AI/ML models also optimize the biomass supply chain, reducing logistics costs.

Predictive Maintenance: AI models (e.g., SVR) predict mixed-ash fusion temperatures to warn of s lagging risks [22]. AI also fuses sensor data (acoustic, visual) to drive "on-demand" intelligent soot blowing, preventing waste and equipment wear.

New Fuel Research: AI (e.g., machine learning combined with DFT) is used to optimize hydrogen/ammonia combustion kinetics and reveal microscopic reaction mechanisms, accelerating the development of next-generation low-NOx combustion technologies [23].

5 AI Empowers CCUS: Breaking Bottlenecks, Reshaping Feasibility

CCUS is considered the ultimate guarantee for deep decarbonization [24], essential for avoiding asset stranding and achieving negative emissions (BECCS) [25, 26]. However, its deployment is stalled by an "Achilles' heel."

5.1 The "Achilles' Heel" of CCUS: High Energy Penalty and Cost

CCUS is trapped in a "vicious cycle." It faces prohibitive capital costs and a massive "energy penalty," consuming 20–30% of a plant's power output [27, 28]. This penalty ($E_{penalty}$) is a direct reduction in net efficiency:

$$E_{penalty}(\%) = \frac{E_{without\_CCUS} - E_{with\_CCUS}}{E_{without\_CCUS}} \tag{5}$$

Furthermore, many flagship projects have failed to meet performance targets [29], deterring investment and preventing the "learning-by-doing" required to lower costs [30]. AI is a catalyst to break this stalemate.

5.2 *Process Optimization: AI-Driven Precise Control*

AI transforms the static, inefficient capture process into a dynamic, adaptive system [31]. For the common amine-based capture process, AI (e.g., ANN and Reinforcement Learning) can dynamically adjust parameters (temperature, flow) to maximize capture rate while minimizing the energy penalty in response to load changes [32]. Simulations based on the Petra Nova project suggest AI could reduce energy consumption by 15% and capture costs by 20–30% [33]. AI-driven predictive maintenance also reduces costly downtime for key equipment like compressors [34].

5.3 *Accelerator for New Materials/Processes*

AI is revolutionizing the discovery of new capture materials. Instead of "trial-and-error," generative AI models can design hundreds of thousands of new candidate molecules (e.g., MOFs) in minutes. Studies confirm AI-driven simulation can speed up new material discovery by up to 500% [33]. AI is fundamentally accelerating the development of next-generation nanomaterials for CCUS [35].

5.4 *"Source-Grid-Load-Storage" Coordination*

AI is also optimizing the transport, storage, and utilization components of the CCUS chain.

5.4.1 Transport and Storage End: Accelerating Site Characterization

Safe geological storage is critical but complex [36]. AI platforms can analyze massive, multi-modal geological data (seismic, logging) to build high-precision 3D models in months, not years. The US DOE's SMART program uses ML-based surrogate models to rapidly predict CO_2 plume behavior, accelerating injection permit evaluations [37]. AI also analyzes microseismic data for real-time risk monitoring.

5.4.2 Utilization End (CCU): from Waste to Value

CCU improves project economics by converting CO_2 into valuable products [38]. AI accelerates the discovery of catalysts for these conversion processes. It is also used to optimize biological fixation (e.g., microalgae) and Enhanced Oil Recovery (EOR), a mature CCU application where AI models (ANN, LSTM) predict and optimize CO_2 injection for maximum oil recovery and sequestration.

6 Cross-Domain Challenges and Integration Paths: Moving Towards Intelligent Autonomous Power Plants

Despite its potential, AI's large-scale deployment faces common bottlenecks, integration challenges, and a clear evolutionary path.

6.1 Common Bottlenecks

1. **Data Issues**: AI models depend on high-quality data, but plant data is often noisy, incomplete, and siloed.
2. **Model Generalizability**: Models trained on one unit may not generalize well to others or to new operating conditions.
3. **The "Black Box" Problem**: The lack of transparency in deep learning models can lead to a lack of trust from operators in high-stakes environments.
4. **System Integration**: Integrating modern AI systems with legacy DCS/SIS control systems is a complex and costly engineering challenge.
5. **AI's Own Carbon Footprint**: The "energy paradox" of using energy-intensive AI models for decarbonization must be addressed.

6.2 The Future of Integration: Digital Twins

The solution to these challenges is system-level collaboration via Digital Twins (DT). A DT is a high-fidelity, dynamic virtual replica of the physical plant [39], integrating IoT data, physics-based models, and AI algorithms (Fig. 3).

The core value of the DT is four-fold:

1. **Breaks Data Silos**: Creates a "single source of truth" for plant-wide optimization.
2. **Provides a Risk-Free Testbed**: New AI-driven control strategies (e.g., from reinforcement learning) can be tested in the virtual environment without risking the physical asset [40].
3. **Global O&M Coordination**: The DT can balance short-term efficiency gains against long-term asset health, finding the true global optimum.
4. **Autonomous Learning**: The DT serves as a training ground for AI agents (RL) to explore and learn control strategies far beyond human experience, enabling a feedback loop where the physical plant continuously improves.

6.3 The Evolutionary Ladder Three-Stage Model

AI empowerment will follow a three-stage evolution:

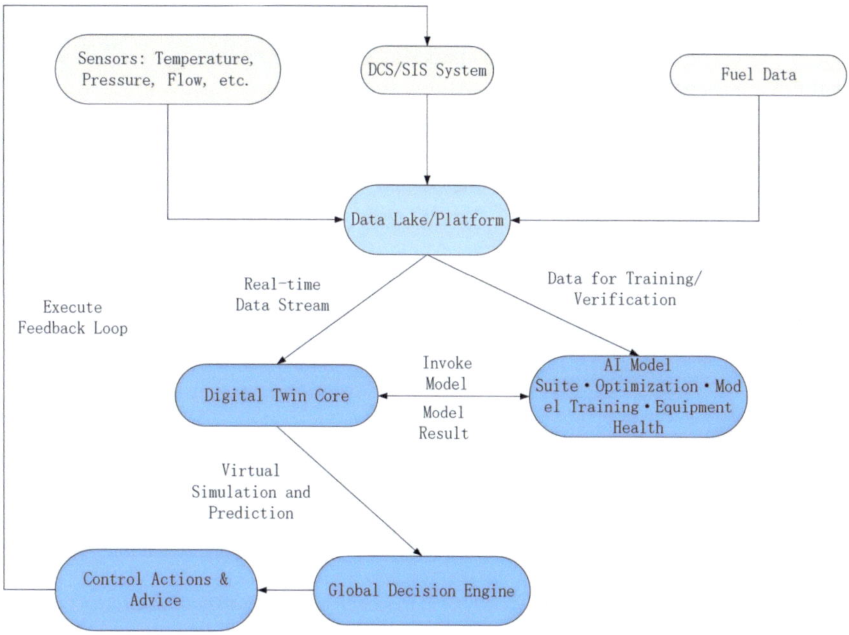

Fig. 3 Digital twin-based integrated optimisation system architecture for intelligent co-firing

1. **Stage 1: Local Optimization**: AI focuses on solving specific, isolated problems, such as single-parameter combustion tuning, NOx emission prediction, or optimizing CCUS absorber energy use.
2. **Stage 2: System Integration**: AI begins to connect different subsystems. Examples include multi-objective combustion optimization (balancing efficiency, NOx, and safety), integrated fuel-supply-chain and co-firing management, and coordinating CCUS operation with plant load.
3. **Stage 3: Autonomous Evolution**: The ultimate goal. A high-fidelity Digital Twin acts as the "brain" of the plant. An AI agent, trained via reinforcement learning, makes plant-wide, autonomous decisions in real-time, balancing efficiency, asset health, and market signals. The system continuously learns and adapts, achieving true autonomous operation from the fuel supply chain to the carbon market.

7 Conclusion

AI is not just an incremental tool but a systemic, transformative force for the low-carbon transition of thermal power. It breaks the inherent "efficiency-environment-safety-cost" dilemmas by converting complex industrial problems into data-driven, optimizable models.

The future lies in the evolution from "point" optimization to "surface" integration, centered on Digital Twins. This paper's "three-stage evolution model" (Local Optimization—System Integration—Autonomous Evolution) provides a structured framework for this transition, moving the industry from "automation" to "autonomy."

However, this vision is not yet fully mature. Significant challenges remain, including weak data infrastructure, the "black box" problem of AI trust, and AI's own carbon footprint. Large-scale success will require a deep, cross-disciplinary fusion of AI with traditional engineering expertise and physics-based models. Only by overcoming these hurdles can AI fully unleash its potential, ensuring that thermal power continues to play its vital "ballast stone" role in a sustainable energy future.

References

1. Intergovernmental Panel on Climate Change (2022) Climate change 2022: mitigation of climate change. Cambridge University Press, Cambridge
2. International Energy Agency (2023) World energy outlook 2023. International Energy Agency, Paris
3. Peifeng N, Yunpeng M, Xinxin Z et al (2019) Research and application of intelligent combustion optimization of coal-fired boiler in thermal power plant based on artificial intelligence technology. Chinese J Intell Sci Technol 1(02):163–170 (in Chinese)
4. Esmaeili M, Moradi H (2023) Robust & nonlinear control of an ultra-supercritical coal fired once-through boiler-turbine unit in order to optimize the uncertain problem. Energy 282:128312
5. Pavan AHV, Vikrant KSN, Ravibharath R et al (2015) Development and evaluation of SUS 304H—IN 617 welds for advanced ultra supercritical boiler applications. Mater Sci Eng, A 642:32–41
6. Hou G, Xiong J, Zhou G et al (2021) Coordinated control system modeling of ultra-supercritical unit based on a new fuzzy neural network. Energy 234:121231
7. Ashraf WM, Uddin GM, Arafat SM et al (2021) Strategic-level performance enhancement of a 660 MWe supercritical power plant and emissions reduction by AI approach. Energy Convers Manage 250:114913
8. Yin L, Zhou H (2024) Modal decomposition integrated model for ultra-supercritical coal-fired power plant reheater tube temperature multi-step prediction. Energy 292:130521
9. Basu P, Butler J, Leon MA (2011) Biomass co-firing options on the emission reduction and electricity generation costs in coal-fired power plants. Renew Energy 36(1):282–288
10. Al-Mansour F, Zuwala J (2010) An evaluation of biomass co-firing in Europe. Biomass Bioenerg 34(5):620–629
11. Hu S, Zhang Y, Ni Y et al (2023) Experimental investigation of combustion and ash deposition characteristics of coal-NH3 cofiring in a 200 kW furnace. Fuel 350:128797
12. Liu W, Zhang H, Wang K et al (2024) The effect of H_2 and NH_3 on the pyrolysis of coal model compounds based on reactive molecular dynamics simulation. J Anal Appl Pyrol 179:106501
13. Zhang J, Elbanna AM, Zhu J et al (2025) Research on the state-of-the-art of efficient and ultra-clean ammonia combustion: From combustion kinetics to engine applications. Appl Energy 391:125886
14. Yao X, Zhou H, Xu K et al (2020) Systematic study on ash transformation behaviour and thermal kinetic characteristics during co-firing of biomass with high ratios of bituminous coal. Renew Energy 147:1453–1468
15. Xing J, Luo K, Wang H et al (2021) A DNS study on temporally evolving jet flames of pulverized coal/biomass co-firing with different blending ratios. Proc Combust Inst 38(3):4005–4012

16. Tan CK, Wilcox SJ, Ward J (2006) Use of artificial intelligence techniques for optimisation of co-combustion of coal with biomass. J Energy Inst 79(1):19–25
17. Wu Z, Gao Z, Li D et al (2021) On transitioning from PID to ADRC in thermal power plants. Control Theory Technol 19(1):3–18
18. Ariemma GB, Sorrentino G, Ragucci R et al (2022) Ammonia/Methane combustion: stability and NOx emissions. Combust Flame 241:112071
19. Wang X, Fan W, Chen J et al (2024) Experimental study of effect of a novel ammonia/coal co-firing mode on NOx emission under high temperature conditions. Renewable Energy 235:121379
20. Liou JL, Liao KC, Wen HT et al (2024) A study on nitrogen oxide emission prediction in Taichung thermal power plant using artificial intelligence (AI) model. Int J Hydrogen Energy 63:1–9
21. Datta A, Gajera B, Saikia M et al (2025) A comprehensive analysis and optimization of coal–biomass mixed fuel for sustainable power generation using the design of experiments and artificial neural network. Clean Technol Environ Policy 27(6):2615–2628
22. Lei M, Han H, Tian X et al (2024) Investigation of ash fusion characteristics on co-combustion of coal and biomass (straw, sludge, and herb residue) based on experimental and machine learning method. Environ Sci Pollut Res 31(6):8467–8482
23. Cao S, Zhang H, Liu H et al (2024) Optimization of kinetic mechanism for hydrogen combustion based on machine learning. Front Chem Sci Eng 18(11):136
24. Liu HJ, Were P, Li Q et al (2017) Worldwide status of CCUS technologies and their development and challenges in China. Geofluids 2017(1):6126505
25. Fajardy M, Chiquier S, Mac DN (2018) Investigating the BECCS resource nexus: delivering sustainable negative emissions. Energy Environ Sci 11(12):3408–3430
26. Weixiang C, Chunpeng W. Carbon capture, utilization and storage (CCUS) in tight oil reservoir. In: Abu Dhabi international petroleum exhibition and conference. SPE, 2023: D011S032R001
27. Leung DYC, Caramanna G, Maroto-Valer MM (2014) An overview of current status of carbon dioxide capture and storage technologies. Renew Sustain Energy Rev 39:426–443
28. Zheng Y, Gao L, He S et al (2023) Reduction potential of the energy penalty for CO_2 capture in CCS. Front Energy 17(3):390–399
29. Mantripragada HC, Zhai H, Rubin ES (2019) Boundary Dam or Petra Nova-Which is a better model for CCS energy supply? Int J Greenhouse Gas Control 82:59–68
30. Al-Sakkari EG, Ragab A, Dagdougui H et al (2024) Carbon capture, utilization and sequestration systems design and operation optimization: Assessment and perspectives of artificial intelligence opportunities. Sci Total Environ 917:170085
31. Wu X, Shen J, Wang M et al (2020) Intelligent predictive control of large-scale solvent-based CO_2 capture plant using artificial neural network and particle swarm optimization. Energy 196:117070
32. Sha P, Zheng C, Wu X et al (2025) Physics informed integral neural network for dynamic modelling of solvent-based post-combustion CO_2 capture process. Appl Energy 377:124344
33. Dhruv N (2024) AI-driven solutions: Quantitative and case study analysis for transforming CCS Technologies [EB/OL]. [2025–07–11]
34. Aminzadeh A, Sattarpanah Karganroudi S, Majidi S et al (2025) A machine learning implementation to predictive maintenance and monitoring of industrial compressors. Sensors 25(4):1006
35. Chen H, Zheng Y, Li J et al (2023) AI for nanomaterials development in clean energy and carbon capture, utilization and storage (CCUS). ACS Nano 17(11):9763–9792
36. Meng L, Zheng J, Yang R et al (2023) Microseismic monitoring technology developments and prospects in CCUS injection engineering. Energies 16(7):3101
37. Siriwardane H, Mishra S, McKoy M et al (2024) SMART–a comprehensive research and development program to demonstrate application of machine learning for supporting CCS deployment. Available at SSRN 5050692
38. Tao WLW (2024) Carbon capture and recycling technology of carbon resources under the target of carbon neutrality. Int J Low-Carbon Technol 19:2693–2701

39. Varapiang T, Rangton N, Nukkhong W et al (2023) Digital twin of biomass/coal co-firing circulating fluidized bed boiler by using computational fluid dynamics simulation. Energy Rep 9:6–9
40. Park SY, Lee C, Jeong S et al (2024) Digital twin and deep reinforcement learning-driven robotic automation system for confined workspaces: A nozzle dam replacement case study in nuclear power plants. Int J Precis Eng Manuf-Green Technol 11(3):939–962

Power Customer Service Dialogue Method Based on Retrieval-Augmented Generation Technology

Donglai Tang, Pu Huang, Qiang Zhang, Bingsen Li, Xi Ding, and Chi Yu

Abstract With the continuous increase in the scale of power customers, traditional power customer service systems can no longer meet customer service needs. To address the issues of poor understanding of user intentions and low logical coherence in power customer service dialogues, a power customer service dialogue method based on retrieval enhanced generation technology is proposed. Firstly, language and text cleaning was carried out to reduce the impact of data noise on the power customer service dialogue model. Secondly, natural language processing is used for semantic analysis to grasp the real demands of power customers. Based on this, retrieval enhanced generation technology is adopted, combined with external data retrieval and the advantages of generative models, to respond to power customer service conversations. Finally, simulation verification was conducted on the conversation data of power customers from a certain provincial power company, and the results showed that the proposed method has better response performance and shorter response time compared to bidirectional recurrent neural networks.

Keywords Search enhanced generation technology · Power customer service · Semantic analysis · Text cleaning · Dialogue

1 Introduction

The power customer service system is a crucial component of the marketing framework of power grid companies. It consists of manual hotlines, online Q&A systems, and other elements, characterized by its broad coverage of power customer services, comprehensive problem-solving capabilities, and user-friendly operation [1, 2]. Among these, manual hotlines are staffed by power professionals who

D. Tang (✉) · P. Huang · Q. Zhang · X. Ding · C. Yu
Aostar Information Technology Co., Ltd., Chendu, China
e-mail: tangdonglai@sohu.com

B. Li
State Grid Information and Telecommunication Group Co., Ltd., Beijing, China

G. Mu et al. (eds.), *Proceedings of the 2nd Electrical Artificial Intelligence Conference, Volume 1*, Lecture Notes in Electrical Engineering 1587,
https://doi.org/10.1007/978-981-95-7932-7_12

address customer inquiries during fixed hours. However, this approach suffers from low communication efficiency, high skill requirements, and a significant demand for personnel [3]. Online Q&A systems facilitate customer interactions through computer-based services such as operating system browsers and WeChat mini-programs. Nevertheless, these automated systems often fail to accurately understand the queries of power customers, resulting in irrelevant responses [4]. Therefore, exploring precise power customer service dialogue technology is of positive significance for improving the quality of customer service.

Many scholars domestically and internationally have conducted extensive research to enhance the accuracy of power customer service dialogues. Reference [5] proposed a power customer service dialogue method based on a Back Propagation (BP) neural network, which predicts customer dialogue intent using the BP neural network and generates dialogue content through association rule extraction. Reference [6] introduced a power customer service dialogue method based on fuzzy spectral clustering, which classifies customer issues via deep learning and responds to them using fuzzy spectral clustering. Reference [7] presented a power customer service dialogue method based on multi-source data fusion, which employs natural language understanding to interpret customer demands and retrieves relevant responses from a knowledge base. Reference [8] proposed a power customer service dialogue method based on knowledge graphs, which analyzes customer dialogue requirements through intent recognition and generates responses using knowledge graphs. These studies demonstrate that various power customer service dialogue methods have achieved notable results. However, these approaches still struggle to deeply understand customer intent, resulting in poor dialogue logic and low response accuracy.

To address the issues of poor user intent understanding and low dialogue logic in power customer service dialogues, this paper proposes a power customer service dialogue method based on Retrieval-Augmented Generation (RAG) technology. First, language text cleaning is performed to eliminate noise in the input text from power customers, followed by text chunking to provide an optimal text extraction strategy. Second, Natural Language Processing (NLP) is employed to identify the true meaning of the customer's dialogue. Finally, the Retrieval-Augmented Generation (RAG) model is used to generate the dialogue response content.

2 Power Customer Service Dialogue Process

The dialogue method for power customer service based on Retrieval-Augmented Generation (RAG) technology is divided into three parts: preprocessing of power customer dialogue text, identification of power customer dialogue intent, and generation of power customer dialogue content. The workflow of the power customer service dialogue method is shown in Fig. 1.

① Preprocessing of Power Customer Dialogue Text

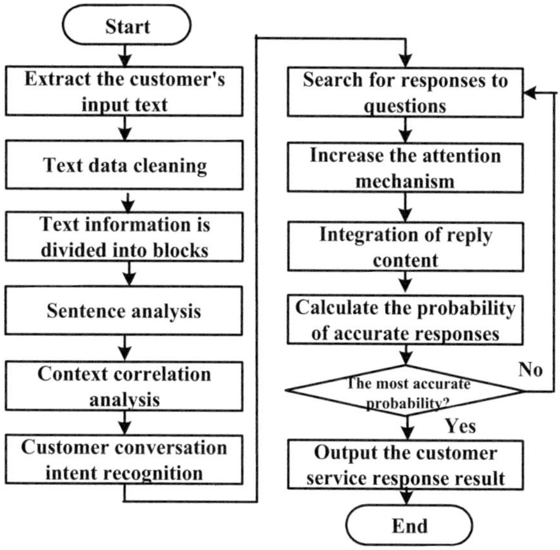

Fig. 1 Power customer service dialogue framework

First, after the power customer service system receives the dialogue information submitted by the customer, the text input by the customer is extracted. Next, through text data cleaning, errors such as incorrect punctuation, repeated words, and non-standard abbreviations are eliminated to prevent noisy data from affecting the model's dialogue responses. Finally, text information is chunked, and long texts are reasonably segmented into paragraphs to facilitate better semantic analysis.

② Identification of Power Customer Dialogue Intent

First, the chunked power customer text information is analyzed at the sentence level to achieve meaning recognition of individual sentences. Next, contextual correlation analysis is performed to correct issues identified in the single-sentence recognition process. Finally, probabilistic prediction is used to identify the customer's dialogue intent.

③ Generation of Power Customer Service Dialogue Content

First, the customer's dialogue question is retrieved to obtain typical response questions from the knowledge base. Next, by incorporating an attention mechanism, the response content is fused, and the probability of response accuracy is calculated. Finally, when the probability of response accuracy is highest, the response result is output.

3 Power Customer Service Dialogue Model

3.1 Preprocessing of Power Customer Dialogue Text

① Extraction of Customer Input Text

The purpose of extracting customer input text in the power customer service system is to collect dialogue information from customers. The extraction methods include text collection and speech collection. For text collection, dialogue text from power customers is obtained through sources such as WeChat mini-programs and customer service websites. For speech collection, text information input by power customers is extracted through speech recognition.

The Bayesian formula is a probability statistical tool that uses conditional probability to describe the relationship between two variables. It is widely used in the field of speech recognition due to its high accuracy [9]. Therefore, the Bayesian formula is adopted for power customer speech recognition.

Let the speech information input by the power customer be denoted as feature vector $\boldsymbol{X}^{\mathrm{IN}} = (\boldsymbol{X}_1^{IN}, \boldsymbol{X}_2^{IN}, \ldots\ldots, \boldsymbol{X}_n^{IN})$, where n is the length of the speech information. Given $\boldsymbol{X}^{\mathrm{IN}}$, the probability of speech recognition text output $P(\mathbf{Y}_i|X^{\mathrm{IN}})$.

$$P(\boldsymbol{Y}_i|\boldsymbol{X}^{\mathrm{IN}}) = \frac{P(\boldsymbol{X}^{\mathrm{IN}}|\boldsymbol{Y}_i)P(\boldsymbol{Y}_i)}{\sum_{j=1}^{n^{\mathrm{P}}} P(\boldsymbol{Y}_j)P(\boldsymbol{X}^{\mathrm{IN}}|\boldsymbol{Y}_j)} \tag{1}$$

where:

n^{P} is the number of complete event groups in the power customer's speech.

$Y_1, Y_2, \ldots, Y_{n^{\mathrm{P}}}$ form a complete event group of the sample space Ω, with $Y_i Y_j = \phi$, $i \neq j$ $(i, j = 1, 2, \ldots, n^{\mathrm{P}})$.

② Text Data Cleaning

The extracted power customer text information may contain noise, such as incorrect punctuation, repeated sentences, and other issues, which can severely affect the accuracy of semantic recognition and the quality of dialogue responses. Therefore, text data cleaning is performed before semantic recognition.

Text data cleaning consists of format processing and redundancy elimination. In format processing, non-text characters and missing data are removed from the power customer text, and the text is converted to a consistent encoding. For redundancy elimination, the Pearson correlation coefficient is used to compare and identify duplicate sentences. Redundant data are deleted to reduce storage space.

The Pearson correlation coefficient is a method for comparing sentence similarity. It is derived from Euclidean distance and uses vector centralization to compute cosine distance, determining whether two sentences are duplicates [10]. The similarity d^{Pccs} between two sentences in the text is calculated as:

$$d^{\text{Pccs}} = \frac{\sum_{i=1}^{n^{\text{F}}} \left(e_i - \bar{e}\right)\left(f_i - \bar{f}\right)}{\sqrt{\sum_{i=1}^{n^{\text{F}}} \left(e_i - \bar{e}\right)^2} \sqrt{\sum_{i=1}^{n^{\text{F}}} \left(f_i - \bar{f}\right)^2}} \tag{2}$$

where:

n^{F} is the number of words used for redundancy elimination.

e_i is the i-th word in the first sentence.

$\bar{e}$ is the benchmark value of words in the first sentence.

f_i is the i-th word in the second sentence.

$\bar{f}$ is the benchmark value of words in the second sentence.

If two sentences are similar, the duplicate sentence is deleted to eliminate redundancy.

③ Text Information Chunking

Efficient text information chunking improves the accuracy of power customer dialogue semantic recognition. Long power customer texts are decomposed into shorter, semantically complete chunks. These chunks adhere to text window limitations while maintaining coherence.

Recursive chunking is an iterative text segmentation method that uses delimiters to hierarchically split power customer text information, iterating until the text is divided into smaller chunks.

Assume the power customer text information x^{EL} contains m^{EL} words, and the k-th word is x_k^{EL}. The word x_k^{EL} is mapped to an n^{EL} dimensional vector space using a standard normal distribution: $x_k^{\text{EL}} \in \mathbf{R}^{n^{\text{EL}}}$. The word vectors are placed into a power customer word embedding matrix Z:

$$Z \in \mathbf{R}^{n^{\text{EL}} \times |B|} \tag{3}$$

where B is the size of the power customer vocabulary.

The power customer word x_k^{EL} can be expressed as:

$$x_k^{\text{EL}} = Zl_k \in \mathbf{R}^{n^{\text{EL}}} \tag{4}$$

where l_k is the size of the power customer text vocabulary in one dimension.

At the start of iteration, the model is unaware of the sentence structure of the power customer text. Therefore, a self-learning tree structure is used for iteration. The objective function for the optimal tree structure of power customer text chunking $f(x_k^{\text{EL}})$ is:

$$f\left(x_k^{\mathrm{EL}}\right) = \underset{y \in A\left(x_k^{\mathrm{EL}}\right)}{\arg\min} \sum_{j \in \varpi(y)} q^{\mathrm{ERR}}\left(c_j\right) \tag{5}$$

where:

$A(x_k^{\mathrm{EL}})$ is the set of all possible tree structures for the power customer text.

y is one tree structure for power customer text chunking.

j is a triple structure in the chunking calculation process.

c_j is the parent node of the triple structure.

$\varpi(y)$ is the retrieval function for the triple structure.

q^{ERR} is the reconstruction error of recursive iteration.

Through iteration, the tree structure for power customer text chunking is optimized.

3.2 Power Customer Dialogue Intent Recognition

Power customer dialogue intent recognition consists of sentence analysis, contextual correlation analysis, and customer dialogue intent identification.

Natural Language Processing (NLP) is a digital processing method for human language and a branch of artificial intelligence focused on voice processing. It can accurately process power customer text information and identify its true intent [11]. Therefore, NLP is used for power customer dialogue intent recognition.

In sentence analysis, the dimensionality of power customer input text is reduced to simplify the analysis. Mutual information is used to measure the correlation between feature words and categories. Related power customer words are clustered, and categories are used to represent the text for dimensionality reduction. The dimensionality reduction formula $f^{\mathrm{MI}}(v_k, H)$ is:

$$f^{\mathrm{MI}}(v_k, H) = \log \frac{P(v_k, H)}{P(v_k)P(H)} = v^{\mathrm{D}} \tag{6}$$

where:

$P(v_k, H)$ is the probability of power customer text feature v_k appearing in class H.

$P(v_k)$ is the probability of v_k appearing in the entire power customer dialogue corpus.

$P(H)$ is the probability of class H appearing in the entire corpus.

v^{D} is the data after dimensionality reduction.

Next, sentence analysis is performed using a text bag-of-words model. The bag-of-words is a collection of specialized phrases for power customer service dialogues. This analysis accurately extracts sentence information from the power customer input text [12]. The current bag-of-words s^{DC} for power customer input text is:

$$s^{\mathrm{DC}} = \frac{1}{\left|o^{\mathrm{C}}\right|}\sum_{j=1}^{c} v_j^{\mathrm{D}}, v_j^{\mathrm{D}} \in o^{\mathrm{C}} \tag{7}$$

where:

c is the number of clusters in the power customer input text.

v_j^D is the feature information of the power customer input text after dimensionality reduction.

o^{C} is the cluster centroid, which serves as the classification center for the bag-of-words.

The cluster centroids are obtained using K-means clustering. First, a random power customer text entry from the bag-of-words is selected as the initial centroid and placed in the clustering set. The Euclidean distance between each word and this centroid is calculated, and words are assigned to the nearest cluster center, forming K clusters. Finally, the reduced-dimensional power customer input text v_j^D is assigned to its respective cluster based on the current input.

Contextual correlation analysis addresses word recognition errors in power customer input text sentence analysis. It effectively corrects deviations in word recognition by considering contextual relationships.

The feature set $\boldsymbol{V}^{\mathrm{Z}}$ for contextual management analysis is:

$$\boldsymbol{V}^{\mathrm{Z}} = \left(\boldsymbol{v}_1^{\mathrm{RAG}}, \boldsymbol{v}_2^{\mathrm{RAG}}, \ldots, \boldsymbol{v}_{n^{\mathrm{TS}}}^{\mathrm{RAG}}\right) \in \gamma \tag{8}$$

where:

n^{TS} is the number of features in the contextual management analysis set.

$\left(\boldsymbol{v}_1^{\mathrm{RAG}}, \boldsymbol{v}_2^{\mathrm{RAG}}, \ldots, \boldsymbol{v}_{n^{\mathrm{TS}}}^{\mathrm{RAG}}\right)$ is the feature of a power customer dialogue word.

γ is the scope of power customer contextual correlation.

Through contextual correlation analysis of power customer input text, the accuracy of sentence recognition is improved.

The information after contextual correlation is used for customer dialogue intent recognition. This intent is then input into the power customer service dialogue content generation module.

3.3 *Power Customer Service Dialogue Content Generation*

The technique for retrieving response questions involves searching the knowledge base for content related to the current customer's dialogue intent and obtaining typical responses. Based on the context of the current customer's dialogue intent, search results are ranked to prioritize the most relevant response information. During retrieval optimization, interactive learning from power customer feedback enhances the understanding of complex queries and improves retrieval performance.

The attention mechanism is a computer simulation of human vision. It focuses the power customer service dialogue model on the task of generating dialogue content, weighting the input data to prioritize the most relevant information for the current dialogue [13]. Therefore, the attention mechanism is used to integrate the retrieved response content.

Retrieval-Augmented Generation (RAG) is a generative model enhanced by external knowledge retrieval. RAG combines external data retrieval with generative models, effectively addressing the lack of diversity in traditional Q&A systems and the issues of uncontrollable output and poor interpretability in generative models [14]. Thus, RAG is adopted for response content retrieval and integration.

The power customer dialogue text z^{OUT} generated by RAG is:

$$z^{\mathrm{OUT}} = \sum_{g} z_{\lambda}(g|x) \prod_{k}^{M} z_{\sigma}(y_k|x, g) \tag{9}$$

where:

g is the input power customer dialogue intent text.

M is the number of generated texts.

y_k is the k-th word generated by RAG.

x is the externally retrieved result.

Z_{λ} is the probability of external retrieval for power customer dialogue.

z_{σ} is the generative probability for power customer dialogue.

During RAG computation, when the probability of response accuracy is optimal, the power customer dialogue text response result is output.

4 Case Study Analysis

4.1 Scenario and Parameter Settings

The proposed retrieval-augmented generation (RAG)-based power customer service dialogue method was validated at a provincial power company's customer service center. Dialogue data was collected through the center's services, resulting in a total of 125,359 power customer service dialogues. Among these, 115,359 dialogues were used as the training set, and the remaining 10,000 were used as the validation set.

The algorithm was deployed on a workstation with the following hardware specifications: an Intel Core i7-10870H processor with a base frequency of 2.2 GHz and a maximum turbo frequency of 5 GHz, 8 cores, 16 threads, 64 GB of RAM, and an integrated Intel UHD Graphics 750 GPU running at 1.3 GHz. The hard drive capacity was 20 TB. The software environment included Windows 11 as the operating system, Python as the model development language, and MySQL as the database.

Parameter settings were as follows: chunk size of 256, word embedding dimension of 200, and a learning rate of 0.02. The proposed method was compared with a bidirectional recurrent neural network (Bi-RNN) [15], which is widely used in power customer service question-answering systems. Therefore, a comparative experiment was conducted between the Bi-RNN and the proposed method.

4.2 *Case Study Execution Analysis*

(1) **Analysis of Training Loss in Power Customer Service Dialogue Model**

Selecting an appropriate number of iterations is crucial for improving model training accuracy. If the number of iterations is too low, the model learns inadequately, resulting in high loss. Conversely, too many iterations lead to overfitting. To validate the training convergence speed of the proposed power customer dialogue model, both the proposed model and the Bi-RNN were trained for 200 iterations. The training loss rates of the power customer service dialogue models are shown in Fig. 2.

As shown in Fig. 2, the proposed RAG-based power customer service dialogue method achieved a loss rate convergence of 0.3% at around 80 iterations, while the Bi-RNN model converged at a loss rate of 0.5% after 110 iterations. Thus, the proposed RAG method achieved higher accuracy with fewer training iterations compared to the Bi-RNN.

(2) **Effectiveness of Power Customer Service Dialogue Model Generation**

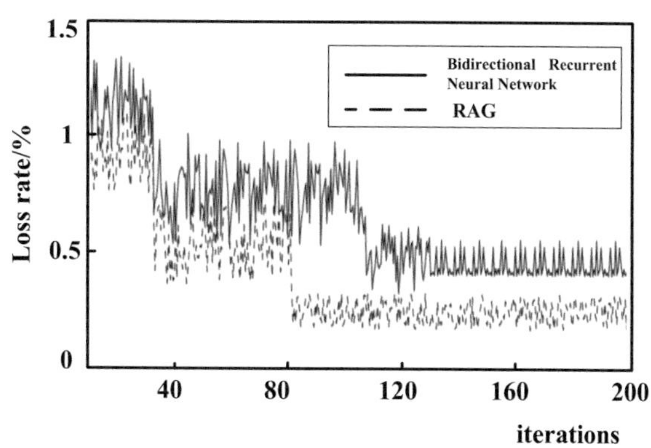

Fig. 2 Training loss rate of power customer service dialogue models

To verify that the proposed RAG-based power customer service dialogue method outperforms other methods, the proposed model was compared with the Bi-RNN and a knowledge graph-based power customer dialogue model. Questions were input into each of the three models, and the responses are compared in Table 1.

As shown in Table 1, compared to the Bi-RNN and knowledge graph methods, the proposed RAG method integrates contextual information, understands the true intent of power users, and thus provides more relevant responses to power customers' questions.

(3) **Performance Analysis of Power Customer Service Dialogue Model**

The performance of the power customer service dialogue model is a core metric for measuring system processing speed. Longer response times lead to poorer customer experiences, while shorter response times improve customer satisfaction. The performance of the dialogue model was evaluated using the response time metric, defined as the duration from when the customer inputs a message to when the system outputs a response.

Fifty texts with lengths of 100, 200, and 300 characters were selected, and the response times of the proposed RAG-based power customer service dialogue method and the Bi-RNN method were compared. The average response times of the power customer service dialogue models are shown in Table 2.

As shown in Table 2, the proposed RAG-based power customer service dialogue method achieved response times of 3.4, 4.3 and 7.7 ms for input lengths of 100, 200, and 300 characters, respectively, with an average of 5.1 ms. These results are significantly lower than those of the Bi-RNN, demonstrating the superior performance of the proposed method.

Table 1 Comparison of power customer dialogue models

Input question	Reply content		
	RAG	Bi-RNN	Knowledge graph
Are there any discounts on electricity bill payments?	Charging on the monthly charging day allows for discounts based on the payment amount	There are discounts on charging days	Discounts available
How many tiers are there in the residential tiered electricity pricing?	Residential tiered electricity pricing has three tiers: within 180 kWh, 180–280 kWh, and above 280 kWh, with different prices for each tier	Three tiers	Three tiers
Which month's electricity bill should I pay?	You should pay last month's electricity bill	Last month's bill	Last month

Table 2 Average response time of power customer service dialogue models

Input character length	Average single response time (ms)	
	Proposed method	Bi-RNN
100	3.40	8.70
200	4.30	17.50
300	7.70	34.20
Average	5.10	20.10

5 Conclusion

To address the issues of poor user intent understanding and low dialogue coherence in power customer service dialogues, a retrieval-augmented generation (RAG)-based power customer service dialogue method was proposed. This method has the following characteristics:

(1) It employs language text cleaning to format power customers' input texts and eliminate redundant information, reducing the impact of data noise on the dialogue model.
(2) It uses natural language processing (NLP) to identify power customers' semantics, enabling a better understanding of their true intentions.

Funding This research was funded by the Chendu Key R&D Program of China grant number 2025-YF08-00109-GX.

References

1. Wu G (2004) Research on Jiangsu electric power customer service system. Jiangsu Electr Eng 23(2):4–8
2. Gao P, Li F, Peng YH et al (2023) Accurate classification method of electric power customers based on Jieba Chinese word segmentation. Hunan Electr Power 43(5):151–154
3. Fu HC, Xu HB (2023) Application of customer value evaluation system based on tag system. Northeast Electr Power Technol 44(4):13–15, 20
4. Tang DL, Qi XK, Chen ZY et al (2022) Power marketing service method based on improved TF-IDF algorithm. Distrib & Util 39(2):22–28
5. Wei DN, Zhang ZL, Yang Y et al (2023) Design of electric power customer service demand prediction model based on BP neural network. Mod Sci Instrum 40(3):11–16
6. Su LW, Chen HY, Tan HC et al (2023) Intelligent Q&A system for electric power customer service based on fuzzy spectral clustering. Inf Technol 47(1):120–125
7. Zheng XH, Zhang JL, Cai Q et al (2023) Intelligent customer service system for electric power enterprises based on multi-source data fusion. Tech Autom Appl 42(5):154–157, 183
8. Shen XJ, Yang HY, Cai Q et al (2023) Research on intelligent customer service system based on power domain knowledge graph. Inform Tech 9:83–90
9. Xu HW, Li ZH (2023) Application of Bayesian formula in genetic counseling. Biol Teach Middle Sch 1:60–62
10. Tang DL, Li Q, Chen YD et al (2023) Identification method of transformer-customer relationship in rural distribution stations based on voltage temporal-spatial aggregation curve. Autom Electr Power Syst 47(19):86–96

11. Weng XZ, Zheng MX, Li CJ et al (2024) Exploration and practice of building an AI operation robot based on NLP technology. Digit Commun World 3:34–37
12. Cheng HB, Wang BA, Chen YR et al (2023) Text clustering based on Gaussian mixture model and adaptive cluster number. J Zhejiang Univ Technol 51(6):602–609
13. Wang LH, Luo ZH, Lin FP et al (2023) C&RM-MAKT prediction algorithm using multi-head attention mechanism. Acta Electron Sin 51(5):1215–1222
14. Tian YL, Wang XX, Wang YT et al (2024) RAG-PHI: retrieval-augmented generation driven parallel humans and parallel intelligence. Chin J Intell Sci Technol 6(1):41–51
15. Tang GL, Xu YF (2024) Design and implementation of intelligent voice detection system for power grid customer service based on machine learning. Microcomput Appl 40(1):217–219, 223

Deep Peak Regulation and Integration of Coal Power with New Energy for New Power System

Jia-Ding Jiang, Xiao-Chao Fan, Jian-Qiao Sun, Zhi-Hao Zhang, Kai Liang, Man-Ting Wan, Xiu-Yu Jiang, and Yi-Xuan Wang

Abstract Research on Deep Peak Regulation of Coal Power and Integration with New Energy Sources for New-Type Power Systems focuses on advanced coal power peak regulation technologies and their integration with renewable energy. With the ongoing transformation of energy structures and rapid development of new energy sources, power systems are facing unprecedented challenges. Xinjiang's abundant wind and solar resources, combined with its large coal-fired power units, provide unique advantages for this research. The team's work primarily covers three key areas: (1) Optimization of Deep Peak Regulation Technology for Xinjiang Coal Power Units. This involves enhancing the flexibility and peak regulation capabilities of coal power units in Xinjiang, particularly through technical upgrades and optimizations for stable operation under low load conditions, rapid start-stop cycles, and response to renewable energy fluctuations. These improvements ensure system stability during high renewable energy integration. Through technological innovation, we aim to boost coal power units' stability under low-load conditions and enable flexible operations. (2) Synergistic Control Between New Energy and Coal Power. This research explores integrated control strategies for renewable energy and thermal power plants, achieving coordinated scheduling and management between new energy sources and coal units to enhance grid flexibility and reliability. (3) Integration of Integrated Energy Systems with Energy Storage Technologies. We investigate the role of energy storage in coal-power integration, design and optimize multi-energy complementary systems, improve grid flexibility, reduce energy losses during peak regulation, achieve low-carbon and efficient energy utilization, and strengthen grid stability through diversified energy complementarity.

Keywords Deep peak regulation · Integration of coal power with new energy · New power system · Xinjiang coal power units

J.-D. Jiang · X.-C. Fan (✉) · K. Liang · M.-T. Wan · X.-Y. Jiang · Y.-X. Wang
The Key Laboratory of Green Hydrogen Production and Storage in Xinjiang, College of Energy Engineering, Xinjiang Institute of Engineering, Urumqi, Xinjiang, China
e-mail: fxc0102@126.com

X.-C. Fan · J.-Q. Sun · Z.-H. Zhang
School of Intelligence Science and Technology, Xinjiang University, Urumqi, Xinjiang, China

G. Mu et al. (eds.), *Proceedings of the 2nd Electrical Artificial Intelligence Conference, Volume 1*, Lecture Notes in Electrical Engineering 1587,
https://doi.org/10.1007/978-981-95-7932-7_13

1 Introduction

The report of the 20th National Congress of the Communist Party of China emphasizes advancing carbon peaking and carbon neutrality in an active yet prudent manner. Grounded in China's energy resource endowment, it advocates a phased approach with orderly transition, implementing carbon peaking actions through systematic planning and step-by-step implementation. The document highlights deepening the energy revolution, enhancing clean and efficient coal utilization, and accelerating the development of a new energy system. Notably, building a new-type power system dominated by renewable energy has become imperative for driving clean energy transformation, supporting carbon peaking goals, aligning with technological advancements, and achieving high-quality development in the power sector while serving the new energy framework [1, 2].

Key pathways include expanding renewable energy integration, improving grid peak regulation capacity, and upgrading traditional coal-fired power plants. The Party and government have prioritized these efforts through multiple policies. The "Guidelines on Enhancing Grid Peak Regulation, Energy Storage, and Intelligent Dispatching Capabilities" mandates that by 2027, regions with high renewable energy penetration and insufficient peak regulation capacity should explore deep peak regulation for coal-fired units, maintaining minimum output below 30% of rated capacity to boost short-term peak response and deep regulation capabilities. The "2024–2025 Energy Conservation and Carbon Reduction Action Plan" requires coordinated efforts in three key areas: energy-saving upgrades, flexibility improvements, and heating system modernization for coal power plants, alongside enhanced renewable energy integration. The "Action Plan for Accelerating the Construction of a New Power System (2024–2027)" outlines nine key special initiatives to be implemented from 2024 to 2027, aiming to achieve tangible progress in building a modern power system. In the coal power peak regulation sector, the plan focuses on clean and low-carbon operations, efficient load adjustment capabilities, rapid load variation, and start-stop peak regulation. These measures will enhance coal-fired units' peak-shaving efficiency and rapid ramping capacity, thereby strengthening their role in ensuring stable electricity supply while facilitating renewable energy integration [3–5].

2 System Composition and Operation

2.1 *Strategic Context in Xinjiang Region*

The "Ten Major Industrial Clusters" in Xinjiang, particularly the "Coal-Electricity-Chemical Industry Cluster" and "New Energy-New Materials", are closely linked to research on "Deep Peak Regulation Integration of Coal Power with New Energy". The "Coal-Electricity-Chemical Industry Cluster" directly contributes to coal power peak regulation by enhancing grid load balancing, facilitating renewable energy integration, and achieving efficient clean utilization of coal resources through modern coal chemical technologies. While new energy development primarily drives coal power peak regulation, advancements in new materials technology—such as advanced energy storage systems and high-efficiency power generation materials—have alleviated this burden. These innovations optimize grid flexibility, store renewable energy, reduce peak regulation demands, and accelerate green energy transition. Xinjiang possesses abundant coal power resources and renewable energy sources (including wind and solar power). As of July 2024, the region's installed new energy capacity reached 80.03 million kilowatts, accounting for nearly half of its total grid capacity. With the advancement of carbon peaking and carbon neutrality goals, the proportion of new energy continues to rise annually, increasing pressure on grid stability. This necessitates urgent solutions: ensuring stable operation of coal-fired units below 30% rated load for deep peak regulation, improving grid flexibility, and achieving coordinated dispatching between coal power and new energy.

2.2 Challenges in Xinjiang Region

While new energy development is booming with large-scale integration of wind and solar power into the grid, these intermittent and unstable energy sources impose higher demands on grid regulation capabilities. With the growing share of renewable energy in China's power mix, the grid urgently requires technological innovations to enhance its regulation capabilities. Coal-fired power units have traditionally served as the backbone for peak load management due to their unique advantages. However, these units face challenges such as low efficiency, high emissions, and limited flexibility during low-load operations, making them inadequate for meeting the massive integration demands of renewable energy. For Xinjiang Province, improving the proactive support and flexible regulation capabilities of wind and thermal power—two primary energy sources—is crucial for ensuring the safe and stable operation of the power system. Moreover, beyond the three main energy sources (thermal power, wind power, and solar power) already integrated into the grid, Xinjiang possesses diverse complementary energy resources. This multi-energy complementarity helps reduce reliance on traditional fossil fuels and mitigate environmental impacts. Against this backdrop, research teams are pooling their expertise to tackle cutting-edge technologies that address the challenges posed by large-scale renewable energy integration.

3 The Overall Research Framework and the Technology Roadmap

3.1 The Overall Research Framework

The overall research framework of this paper focuses on ensuring coal-fired power units can operate stably below 30% rated load while achieving deep peak regulation—a critical challenge requiring urgent solutions. This study explores the impacts of high renewable energy integration in Xinjiang, clarifying their interrelationships and revealing the specific patterns of supply fluctuations caused by renewable energy penetration on grid peak regulation. For regions with high renewable integration, we investigate technical pathways for coal-fired units to operate below 30% rated load under safe conditions. Simultaneously, based on current wind power integration status, we develop coordinated control methods between renewables and coal plants. This approach ensures deep peak regulation while maintaining grid frequency stability and system coordination, achieving seamless integration and safe operation of Xinjiang's renewable and thermal power generation. Finally, through integrating energy storage technologies into hybrid systems, we enhance coal-renewable integration efficiency, reduce peak regulation costs, and improve overall system flexibility and stability, ultimately enabling low-carbon, efficient operations. The expected research outcomes will provide new methodologies, technical support, and management strategies for optimizing Xinjiang's grid peak regulation capabilities. The overall research framework is illustrated in Fig. 1.

3.2 The Technology Roadmap

The innovation team of deep peak regulation and new energy integration for the new power system focuses on the core tasks of deep peak regulation of coal power units in Xinjiang, coordinated control of new energy and coal power, and integration of comprehensive energy system and energy storage technology. The proposed technical route is shown in Fig. 2.

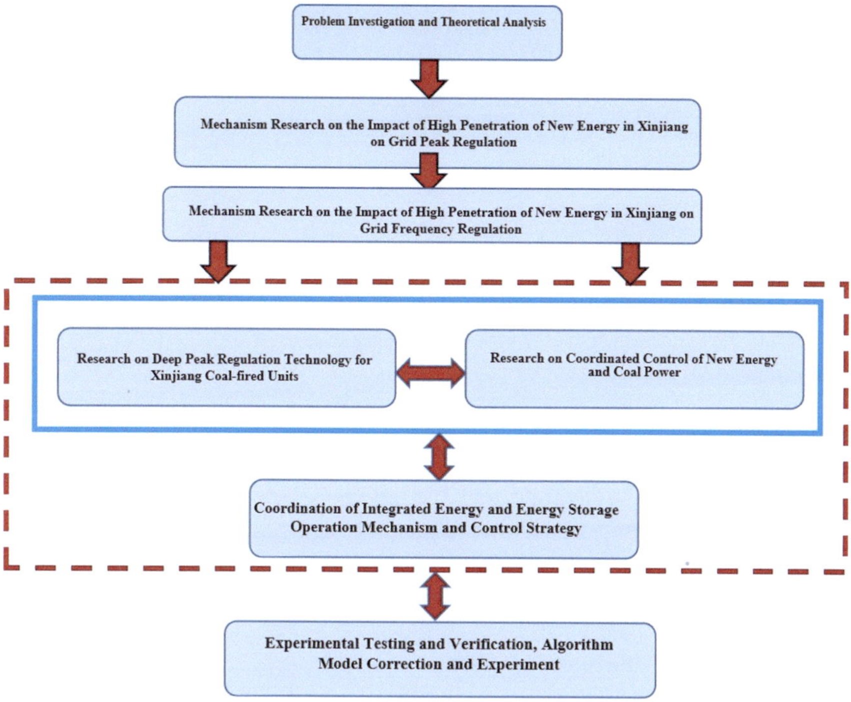

Fig. 1 The overall conception

4 Research on System Performance

4.1 Study on the Influence Mechanism of High Penetration Rate of New Energy in Xinjiang on Peak and Frequency Modulation of Power Grid

By collecting Xinjiang's renewable energy generation data (including wind and solar power), existing grid load data, and peak regulation resource data, we preprocess the data through cleaning and normalization. Statistical methods and machine learning models are employed to analyze the volatility and characteristics of renewable energy generation. We establish models to simulate grid peak regulation demands and capabilities under different renewable energy penetration rates. The study examines how increased peak regulation demands from renewable grid integration affect existing resources, including energy storage facilities and thermal power unit flexibility. Using simulation software, we model the impact of renewable energy supply fluctuations on grid frequency stability. Additionally, we evaluate the response efficiency of grid frequency regulation strategies and resources at varying renewable integration levels. Through mathematical modeling and simulation platforms, this research simulates

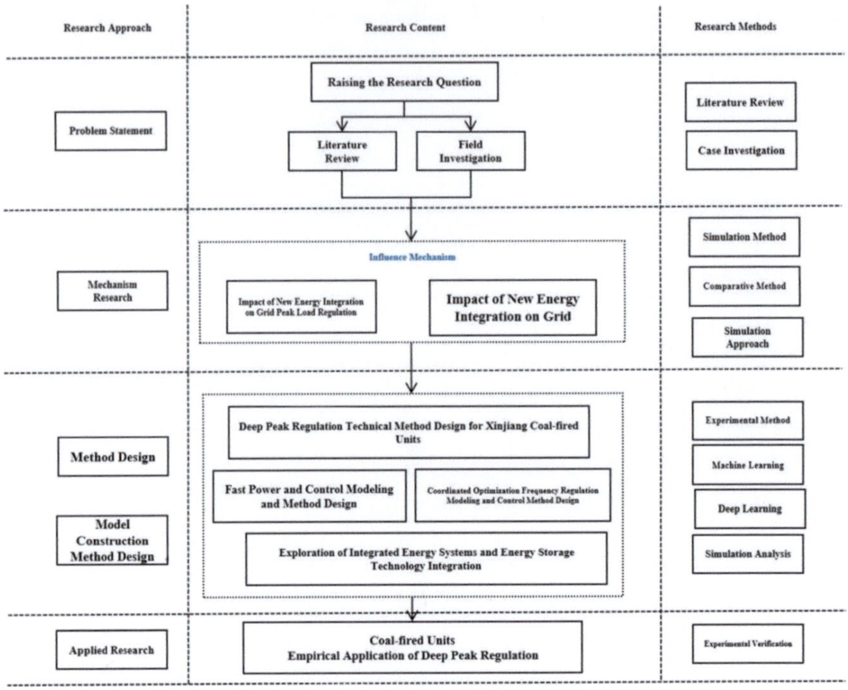

Fig. 2 Schematic diagram of the technical route to be adopted in this paper

the effects of renewable energy integration ratios on grid peak regulation capacity and frequency stability. Specific research content is illustrated in Fig. 3.

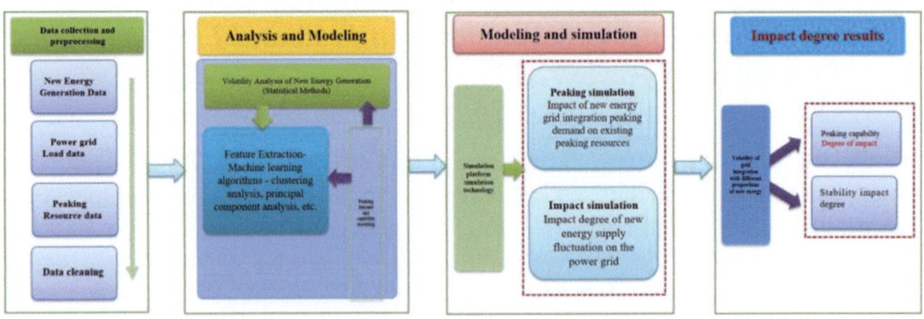

Fig. 3 Research content on the influence mechanism of high penetration rate of new energy in Xinjiang on power grid peak regulation and frequency modulation

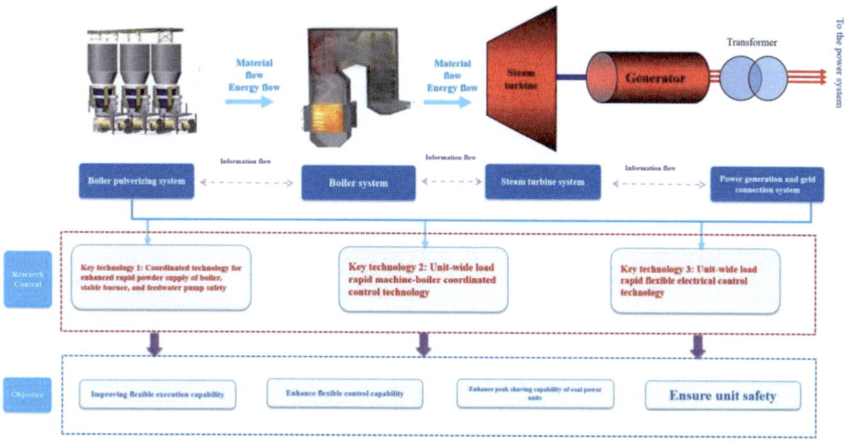

Fig. 4 Research on deep peak regulation technology of coal power units in Xinjiang

4.2 Study on Optimization of Deep Peak Regulation Technology of Coal Power Units in Xinjiang

This research focuses on developing advanced peak-shaving strategies for coal-fired power units in Xinjiang's renewable energy integration environment, where high proportion of new energy sources are connected to the grid. The study optimizes system characteristics to enhance response speed and flexibility, thereby improving peak regulation capabilities and stability. Key technologies include rapid load adjustment mechanisms and thermal system optimization, with particular emphasis on stabilizing low-load boiler operations and enhancing efficiency control. Critical equipment improvements encompass optimized combustion regulation, steam temperature management, feedwater control, turbine performance optimization, and enhanced electrical control systems. Additionally, the research explores coordinated boiler-turbine control systems for flexible peak-shaving under high renewable integration, along with methods to boost the flexibility of supercritical coal-fired units. The main research components are illustrated in Fig. 4.

4.3 Research on Synergistic Control of New Energy and Coal Power Based on Xinjiang's

High-Proportion Wind Power Integration Status, the research team will conduct collaborative generation technology studies between wind power and thermal power. This includes developing rapid power and frequency regulation control methods for wind farm clusters, exploring how advanced control strategies enable wind farms to achieve swift grid response in power and frequency management. The project focuses

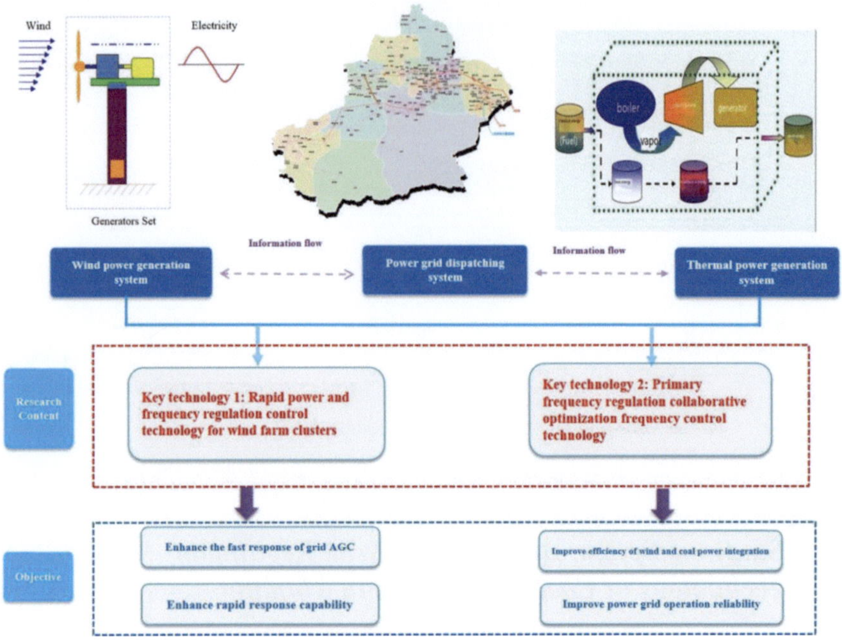

Fig. 5 Research content of coordinated control of new energy and coal power

on optimizing primary frequency regulation (PFR) coordination between thermal power units by considering the coupling characteristics of wind and thermal power frequency modulation. A comprehensive framework will be established to develop coordinated control strategies that ensure effective synchronization between wind and thermal power units during frequency adjustment. Additionally, the research aims to design integrated operation mechanisms and control strategies for deep integration of new energy and coal power, explore synergistic operational models, and enhance grid peak-shaving capacity through coordinated operation of renewable energy (particularly wind power) and coal power. These efforts will improve power system flexibility and reliability. Key research components are illustrated in Fig. 5.

4.4 *Research on Integration of Energy Storage Technology and Integrated Energy System*

Firstly, through system modeling and simulation technologies, we construct integrated energy system models combining wind power, photovoltaic, coal power, and energy storage to design and optimize energy flow and dispatching schemes. By considering regional characteristics and load patterns, we determine optimal multi-energy configurations to enhance system efficiency and energy utilization rates. This

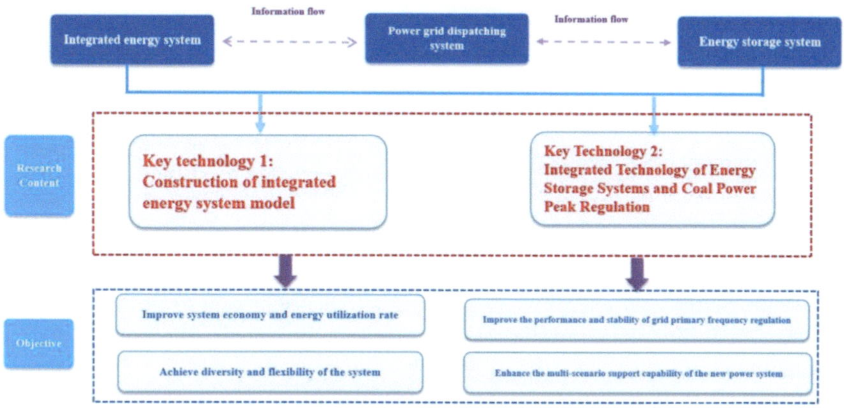

Fig. 6 Research content of integrated energy system and energy storage technology

approach enables diversified energy distribution and complementary integration, achieving both system diversity and flexibility. Secondly, focusing on the performance features of different energy storage technologies, we investigate their applications in coal-power and renewable energy integration. Our research explores peak shaving and valley filling capabilities in high-proportion renewable systems, while developing integrated technologies between energy storage and coal power peaking. To meet the needs of new-type power system planning, we prioritize rapid response and flexible dispatching techniques for energy storage systems. Key studies include active participation strategies for energy storage in system inertia management and primary frequency regulation (PFR). These efforts address renewable energy volatility and uncertainty, improve grid stability through enhanced PFR performance, and boost multi-scenario support capabilities for modern power systems. The main research content is illustrated in Fig. 6.

5 Conclusions

(1) Deep Peak Regulation Technology

To address challenges in traditional thermal power units during deep peak regulation, the research team has implemented a series of technological innovations. These include adopting advanced combustion control technologies and improving stability and efficiency in low-load boiler operations, thereby enhancing peak regulation capabilities and operational efficiency. The team has developed optimized control systems for boiler stability and efficiency under low-load conditions, as well as coordinated optimization mechanisms for coal-fired power plants to adapt to high-proportion renewable energy integration.

(2) New Energy and Power System Integration Technology

In response to the integration of high-proportion renewable energy sources, the team conducts research on wind-thermal power coordination generation technology, developing rapid power adjustment methods and collaborative frequency optimization control strategies for wind farms in regions with abundant renewable energy.

(3) Synergistic Peak Regulation Technology for Renewable Energy and Energy Storage Systems

To overcome limitations in peak regulation caused by large-scale renewable grid integration, the team investigates coordinated peak regulation techniques between renewable energy and energy storage systems. This approach leverages the rapid response capability of energy storage systems to compensate for the intermittent and fluctuating output characteristics of renewable energy sources.

Acknowledgements This research forms part of a study program supported by the National Natural Science Foundation of China (72361033, 52266018), General Project of Science and Technology Department of Xinjiang Uygur Autonomous Region (2022D01A241), Doctoral Start-up Fund Project of Xinjiang Institute of Technology (2023XGYBQJ01). Xinjiang Tianshan Talents -Project for Outstanding Young Scientific and Technological Talents (2022TSYCCX0051), Major Science and Technology Projects of Xinjiang Uygur Autonomous Region in 2024 (2024A01005-1). The authors would most sincerely like to convey their gratitude to the above-mentioned funding agencies for their supportive efforts.

References

1. Mingjiang D, Bo M, Yan L et al (2022) Green transformation pathways for Xinjiang's energy system under the 'Dual carbon' goals. J Nat Resour 37(05):1107–1122
2. Bao J, Li X, Hu Q et al (2024) Exploring temporal-spatial patterns of carbon emissions in Xinjiang's energy consumption and energy structure adjustment pathways. Arid Reg Res 41(03):490–498
3. Yue W, Qian Y, Rui Z (2024) Analysis and research on key coal quality characteristics and utilization approaches of major coalfields in Xinjiang. China Coal 50(04):97–104
4. Shengping L (2025) Research on coal power flexibility retrofitting and synergistic optimization with new energy under the "dual carbon" goals in Northwest China. J North China Electr Power Univ (Soc Sci Ed) 03:58–66
5. Yuan H, Chao W, Yujie Z et al (2024) Current status and synergistic development pathways of coal power and new energy technologies. New Power Syst 2(04):386–405. https://doi.org/10.20121/j.2097-2784.ntps.N20240011

Physics-Informed Neural Networks for Solving Static Magnetic Problems

Youkang Hu, Yuqiu Sun, Shenglei He, Tao Xu, Lin Qin, and Wei Xu

Abstract This work proposes an unsupervised Physics-Informed Neural Network (PINN) framework with Fourier feature encoding to solve partial differential equations (PDEs) describing magnetic field distributions under current excitation. A three-output network is designed to predict magnetic vector potential and magnetic field components, addressing media discontinuities. Two fundamental benchmark models are constructed to demonstrate the effectiveness of the method. Additionally, a relay model is introduced for validation, confirming the framework's accuracy and generalization in practical multi-domain electromagnetic problems.

Keywords Static magnetic field computation · Fourier features · Unsupervised learning · Physics-Informed Neural Networks (PINNs)

1 Introduction

The application of AI for Science is now extending beyond conventional areas such as computer vision and speech processing to address fundamental challenges in scientific computing, notably the numerical solution of partial differential equations (PDEs) that underpin physical system modeling. Traditional discretization-based approaches—including the Finite Element Method (FEM) and Finite Difference Method (FDM)—incur prohibitive computational costs when resolving fine-scale features through mesh refinement [1–4].

Neural network approaches bypass the problem of domain discretization. The universal approximation theorem provides a theoretical foundation, indicating that a sufficiently wide network can approximate any Borel-measurable function, including

Y. Hu · Y. Sun · S. He · T. Xu · W. Xu (✉)
School of Electrical Engineering, Southeast University, Nanjing, China
e-mail: weixu@seu.edu.com

L. Qin · W. Xu
School of Automotive Engineering, Tongji University, Shanghai, China

G. Mu et al. (eds.), *Proceedings of the 2nd Electrical Artificial Intelligence Conference, Volume 1*, Lecture Notes in Electrical Engineering 1587,
https://doi.org/10.1007/978-981-95-7932-7_14

solutions to PDEs [5–7]. Traditional neural network methods are primarily data-driven, requiring large datasets of paired PDE parameters and precise solutions [8]. This dependence brings limitations in terms of data availability, solution accuracy, and inherent non-interpretability (the "Black box" nature). Physics-Informed Neural Networks (PINNs), pioneered by Raissi et al. in 2019, directly embed physical laws into the network's loss function [4]. This paradigm shift reduces reliance on large datasets, improves physical interpretability, leverages automatic differentiation, and offers a mesh-free solution framework. PINNs have demonstrated significant potential in various scientific and engineering applications [9–11].

This paper focuses on magnetostatic partial differential equations, employing Physics-Informed Neural Networks to investigate and solve magnetic field distributions under constant current excitation. The main contributions of this research are outlined as follows: (a) Enhancing Network Expressiveness through Fourier Mapping: The network's input features are transformed from the spatial domain to the frequency domain, enhancing both prediction accuracy and convergence. (b) Three-output network architecture: A three-branch output architecture is designed where Maxwell's equations are structurally embedded at the output layer through dimension adaptation. This configuration inherently compensates for medium parameter variations, reducing their destabilizing influence on training convergence.

2 Methodology

2.1 Physics-Informed Neural Networks (PINNs)

Physics-Informed Neural Networks is a class of deep learning models that seamlessly integrate physical laws into neural network training, enabling mesh-free simulation and data-physics fusion. For magnetostatic problems, this approach leverages Maxwell's equations as foundational constraints.

2.2 Magnetic Field Formulation

Under a steady current excitation [12], Ampère's law in Maxwell's equations leads to Eq. (1).

$$\nabla \times \boldsymbol{H} = J \tag{1}$$

where H is magnetic field intensity, J is current density. The magnetic flux density B is related to the magnetic field intensity through the constitutive relation [6], as shown in Eq. (2).

$$\boldsymbol{B} = \mu_0 \mu_r \boldsymbol{H} \tag{2}$$

where μ_0 is the permeability of vacuum, μ_r is the relative permeability.

In addition, Gauss's law for magnetism indicates that the magnetic flux density is divergence-free, as expressed in Eq. (3).

$$\nabla \cdot \boldsymbol{B} = 0 \tag{3}$$

According to the Helmholtz decomposition theorem any divergence-free vector field can be expressed as the curl of another vector field. Thus, define a vector potential A such that $\boldsymbol{B} = \nabla \times \boldsymbol{A}$, Combining Eqs. (1), (2), and (3), obtain Eq. (4), which is the target equation to be solved. Moreover, the vector magnetic potential A is the primary physical quantity to be determined.

$$\nabla \times \left(\frac{1}{\mu_0 \mu_r}\right) \nabla \times \boldsymbol{A} = J \tag{4}$$

To solve Eq. (4), appropriate boundary conditions must be specified. At the surface of the magnetic body with normal vector n, the normal component of the magnetic flux density and the tangential component of the magnetic field are continuous.

$$\begin{aligned} \boldsymbol{A}_1 &= \boldsymbol{A}_2 \\ H_{1t} &= H_{2t} \\ B_{1n} &= B_{2n} \end{aligned} \tag{5}$$

The reference magnetic vector potential at the outer boundary is enforced by the condition:

$$\boldsymbol{A} = 0 \; and \; H_t \; or \; H_n = 0 \tag{6}$$

2.3 Architecture

Independent neural networks are assigned to regions with distinct physical characteristics, and the consistency between them is maintained by applying physical boundary conditions at the interfaces.

As shown in Fig. 1, demonstrates how different networks are organized and coupled via boundary conditions.

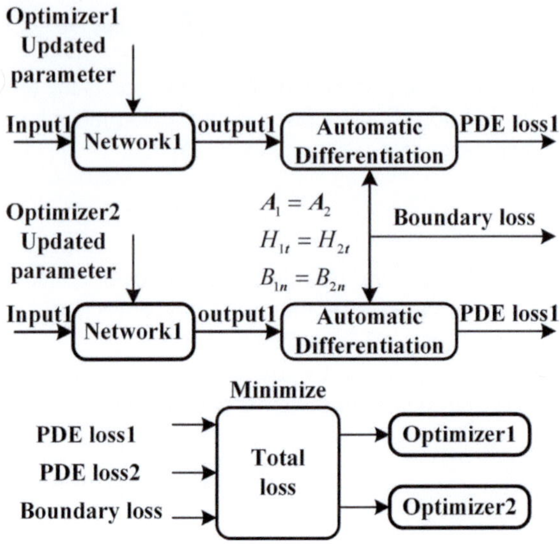

Fig. 1 The solution architecture

2.4 Fourier Feature

As shown in Fig. 2, fourier feature embedding can be regarded as a non-trainable layer that can be incorporated into any Physics-Informed Neural Network (PINN) architecture.

Specifically, the Fourier mapping layer is used to convert the input feature from the spatial coordinate domain to the frequency domain of Fourier space, so as to assist the network to obtain more input details, improve the network's learning ability for high-frequency information with Fourier series characteristics, accelerate the convergence speed of the network model and improve the output fineness [6, 13].

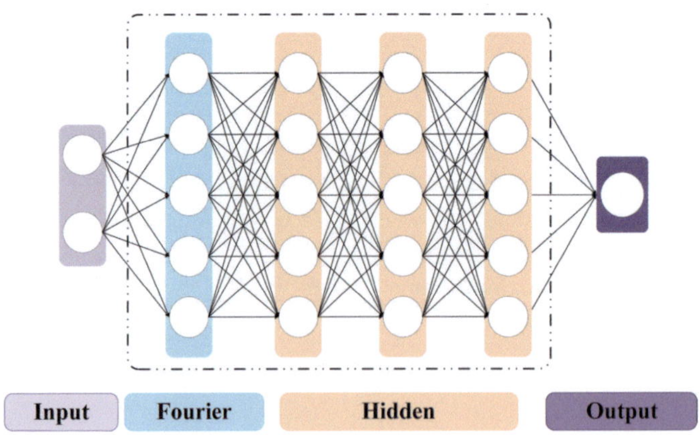

Fig. 2 The Network with Fourier feature

In this transformation process, the Fourier mapping layer participates in the data feature mapping, and the randomly sampled two-dimensional input features are transformed by mathematical transformation, and finally each spatial coordinate is transformed into the frequency domain representation.

The output of the network is obtained by forward propagation, and the required partial derivatives of each order can be calculated with the help of the automatic differentiation mechanism of Pytorch, so as to explicitly construct the total loss function of PINNs. Based on the loss function, the optimizer performs back propagation to realize the iterative update of network parameters [14, 15].

For solving the Laplacian example in a single region, the total loss function of the network is composed of the residual term of the partial differential equation and the boundary loss [16, 17]. Since all the boundaries belong to the first type of boundary, in the forward output of the network, the corresponding output of the boundary does not participate in the automatic differentiation process, and random sampling and uniform sampling are performed on the solution region and the boundary respectively. The spatial characteristics of irregular distribution in the region change dynamically with the update of network parameters [18, 19].

3 Numerical Studies

3.1 Homogeneous Medium

As shown in Fig. 3, this case considers the case with uniform magnetic permeability parameters.

At the air-copper interface, tangential and normal satisfy continuity conditions according to Flux continuity theorem. The outer boundaries (left, bottom, and right) are modeled as ideal magnetic conductors, while the top boundary enforces a Dirichlet condition.

The network training parameters are summarized in Table 1.

The neural network predictions demonstrate strong agreement with commercial FEM solutions (ANSYS Maxwell v2021 R2), as quantitatively compared in Fig. 4 through field distributions and error metrics.

According to the comparison results, the prediction of electromagnetic field distribution by neural network is highly consistent with the results of finite element calculation in the overall distribution characteristics. In order to further improve the accuracy of the evaluation, this paper extracts the position of the boundary and interface in the finite element simulation, collects the space coordinate points and the corresponding vector magnetic potential calculation results, and uses the extracted space coordinates as the input features of the network to predict. So as to realize point-by-point comparison and error analysis, and intuitively reveal the deviation of the key area. The continuity of the vector magnetic potential has been verified in the above subsection, and the results show that the network has high consistency in the interface

Fig. 3 Numerical example setting (units: cm)

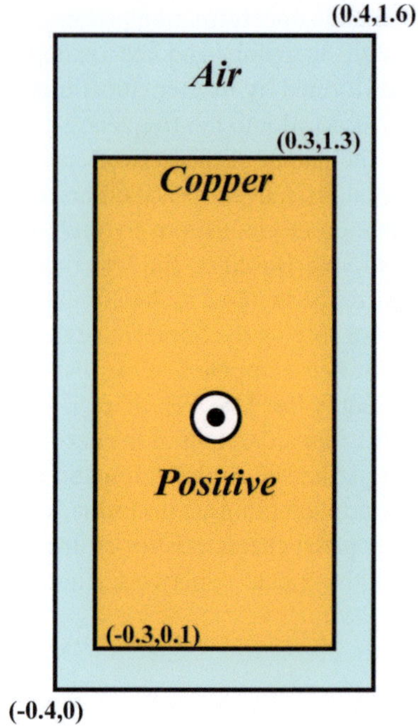

Table 1 Network training hyperparameters

	Air	Copper
μ_r	1	1
J	0	2.5A/mm^2
Inputs	$[800 \times 2]$	$[1000 \times 2]$
Neurons	256	256
Optimizer	Adam	Adam
Hidden layers	3	3
Training steps	5×10^4	5×10^4
Learning rate	1×10^{-4}	1×10^{-4}

position prediction results. Considering that the vector magnetic potential is affected by the current density, the prediction results of the interface position are dominated by the current source area network, and the boundary position is controlled by the air area network.

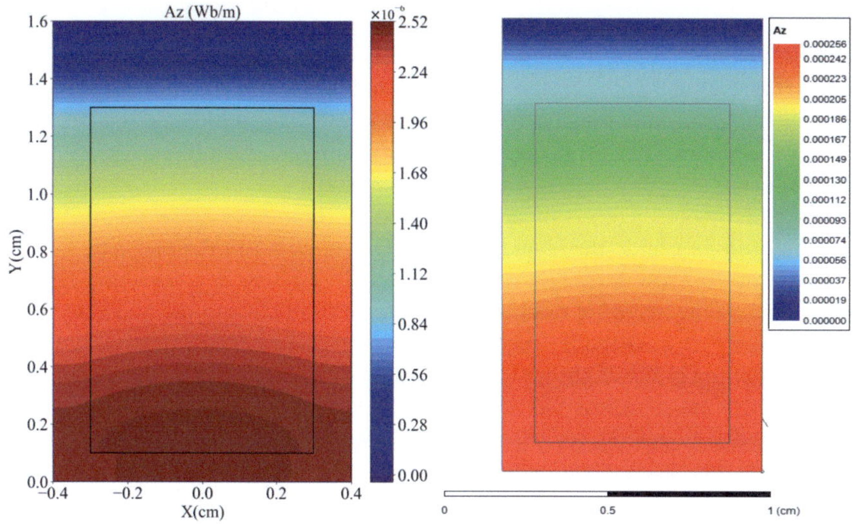

(a) Network (b) FEM

Fig. 4 The vector potential

3.2 *Discontinuous Medium*

Furthermore, the numerical prototype in Fig. 4 demonstrates a high-contrast permeability distribution, with the soft magnetic core ($\mu_{r\_steel} = 3000$) adjacent to non-magnetic regions ($\mu_{r\_copper} = 1$), creating challenging interface conditions for field computation (Fig. 5).

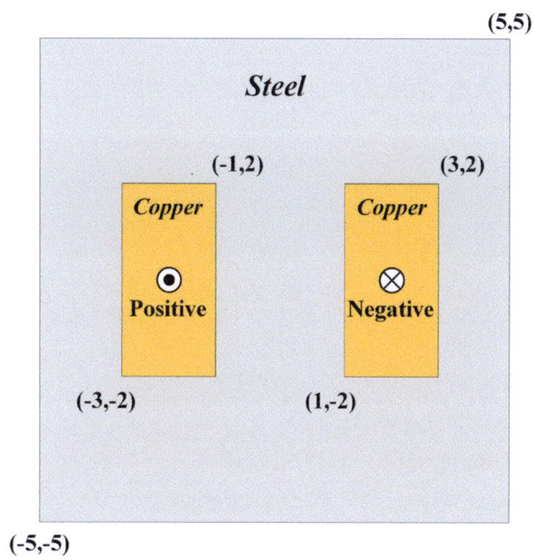

Fig. 5 Numerical example setting (units: m)

The significant challenge posed by the material property contrast at the copper/high-permeability interface, where relying solely on the magnetic vector potential to derive H and B continuity constraints impedes neural network training, the approach of simultaneously outputting the three primary physical quantities and transforming the governing equations using their curl relationships and constitutive laws.

In this case study, the interface between copper and silicon steel maintains satisfaction of the magnetic flux continuity condition. At the outer boundary, the magnetic flux lines run parallel to the boundary edge. This implies that the magnetic field is entirely confined within the silicon steel material, resulting in a vanishing normal component of the magnetic field intensity ($H_n = 0$) relative to the boundary. The corresponding network training parameters are detailed in Table 2.

As evidenced in Fig. 6, the network predictions exhibit excellent agreement with finite element results across the computational domain. The primary discrepancies are confined to singularity regions, where field discontinuities occur.

The network shows good approximation ability in the overall prediction, and can restore the finite element calculation results more accurately. However, in the complex physical environment where current excitation and multi-medium distribution coexist, the prediction accuracy in the local region slightly decreases. The main reason is that the existence of magnetic conductive materials leads to the significant jump of magnetic permeability between different media, and then causes the dramatic spatial gradient change of vector magnetic potential at the boundary, which increases the difficulty of neural network to learn the characteristics of this region. In addition, corner positions in the geometry are often accompanied by abrupt changes in boundary conditions, which become concentrated regions of abrupt changes in the local field distribution under the coupling effect of the current source and the magnetic conductor, thus aggravating the non-smoothness of the solution. Despite these local challenges, the network as a whole can still accurately describe the spatial distribution characteristics of the vector magnetic potential, which verifies the effectiveness of the constructed model.

Table 2 Network training hyperparameters

	Steel	Positive	Negative
μ_r	3000	1	1
J	0	1A/m^2	-1A/m^2
Inputs	$[800 \times 2]$	$[300 \times 2]$	$[300 \times 2]$
Outputs	A_z, H_x, H_y	A_z, H_x, H_y	A_z, H_x, H_y
Neurons	256	256	256
Optimizer	AdamW	AdamW	AdamW
Hidden layers	3	3	3
Training steps	6×10^4	6×10^4	6×10^4
Learning rate	Cosine annealing schedule	Cosine annealing schedule	Cosine annealing schedule

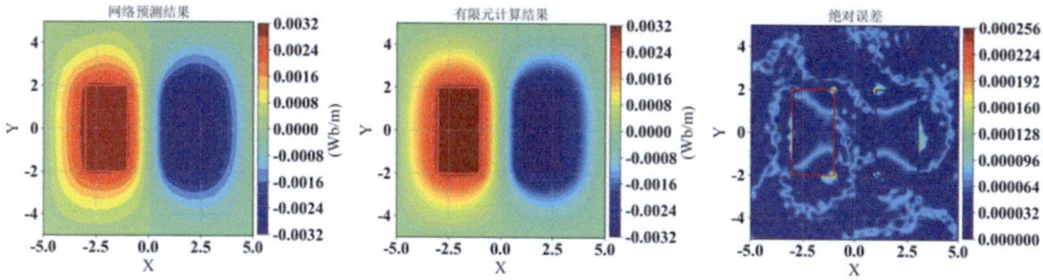

(a) The vector potential

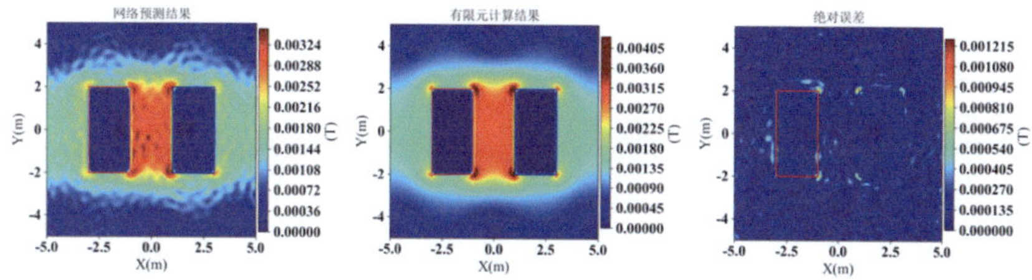

(b) Magnetic flux density

Fig. 6 Predicted results by the PINN model compared with FEM solutions

4 Application to Relay Model

4.1 Relay Model

The synthesized electromagnetic relay model resulting from combined methodologies of above to frameworks is presented in Fig. 7.

Fig. 7 Relay model

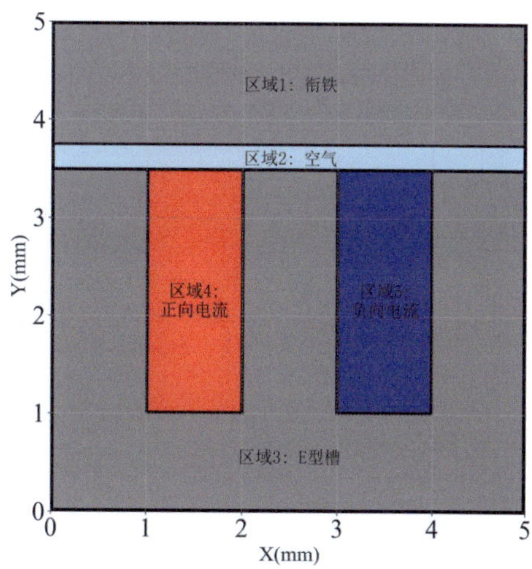

Table 3 Neural network architecture parameters

Parameters	Value
Neurons	256
Optimizer	AdamW
Hidden layers	4
Training steps	1.5×10^5
Learning rate	Cosine annealing schedule

In this numerical experiment, encompassing all case studies from Section III. NUMERICAL STUDIES, the computational domain depicted in Fig. 7 is decomposed into five distinct subdomains. Interface conditions are rigorously configured according to the corresponding material properties and excitation settings. The neural network architecture parameters are detailed in Table 3.

With the increase of the complexity of the permeability distribution in the solution area, in order to enhance the expression ability of the model for complex physical information, a deeper network layer is introduced in the network structure, and the frequency parameter in the Fourier feature map is extended to improve its representation ability under high-frequency features. However, the increase of network depth and the expansion of network number also significantly improve the optimization difficulty of model training. In order to ensure the convergence and stability of training, this paper appropriately extends the training period and increases the number of iterations.

Figure 8 demonstrates the predictive accuracy of the model. The network successfully captures global spatial distributions of magnetic fields and media, with errors predominantly concentrated in air regions and radiating from current source centers. These zones exhibit strong field gradients where error inhomogeneity correlates with the dual-subnetwork architecture implemented during training.

As shown in Table 4, the mean relative error (MRE) and maximum relative error (MaxRE) of each subnetwork are reported.

The electromagnetic force was calculated based on the magnetostatic field results using the Maxwell stress tensor method, as given by Eq. (7), and the corresponding results are summarized in Table 5.

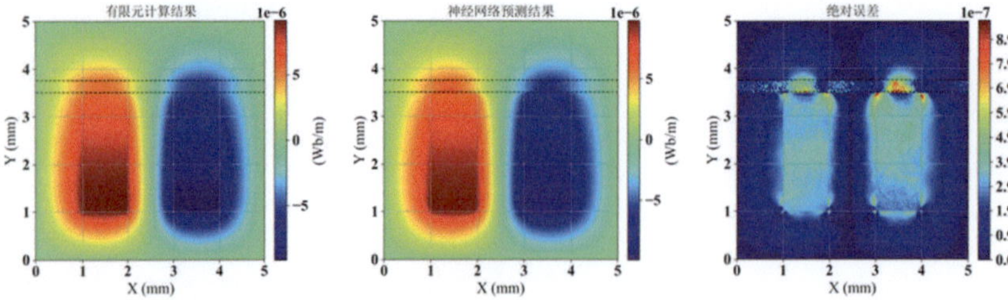

Fig. 8 Relay model

Table 4 Mean relative error (MRE) and maximum relative error

	Yoke	Air	Positive	Negative	E-shaped
MRE [%]	3.90	13.04	4.48	4.58	3.61
MaxRE [%]	10.38	45.05	15.52	19.11	43.32

Table 5 Electromagnetic force results

	FEM	Network
$\boldsymbol{F_x}$	0	0.004744
$\boldsymbol{F_y}$	0.04553	0.04293
$\boldsymbol{F}$	0.04553	0.04319

$$\begin{cases} F_x = \int_l \frac{B_y \cdot B_x}{\mu_0} dl \\ F_y = \int_l \frac{B_y^2 - B_x^2}{2\mu_0} dl \end{cases} \tag{7}$$

The experimental results show that the proposed PINNs solution framework performs well in the relay magnetic field distribution prediction as a whole, and can basically achieve accuracy and physical consistency. Although there are some differences between the electromagnetic force calculation results and the finite element method, the overall trend remains the same, which verifies the engineering applicability of the proposed method under unsupervised conditions.

5 Conclusion

In this work, we developed a Physics-Informed Neural Network (PINN) to predict magnetic field distributions under steady current excitation in media with varying magnetic permeability. We designed the network with a Fourier feature mapping layer using non-trainable parameters and implemented a three-output architecture. By explicitly embedding Maxwell's equations into the loss function, we enabled the network to perform unsupervised learning without relying on labeled data.

Despite its effectiveness, our method presents certain limitations, including long training times and relatively large prediction errors near singularities.

Acknowledgements This work was supported in part by the Shanghai Municipal Education Commission under the "Program for AI-Empowered Reform of Scientific Research Paradigms and Disciplinary Advancement", in part by Jiangsu Frontier Technology Research and Development Program "Robot with high-performance integrated integrated electric drive system technology research and development" under Grant BF2024064, in part by National Natural Science Foundation of China under Grant 52275009 and Grant 52207038.

References

1. Cybenkot G (1989) Approximation by superpositions of a sigmoidal function. Math Control, Signals, Syst (MCSS) 2:303–314
2. Raissi M, Yazdani A, Karniadakis GE (2016) Hidden fluid mechanics: Learning velocity and pressure fields from flow visualizations. Science 367(6481):1026–1030. CONFERENCE 2016, LNCS, vol 9999, pp 1–13. Springer, Heidelberg
3. Funahashi KI (1989) On the approximate realization of continuous mappings by neural networks. Neural Netw 2(3):183–192
4. Raissi M, Perdikaris P, Karniadakis GE (2019) Physics-informed neural networks: A deep learning framework for solving forward and inverse problems involving nonlinear partial differential equations. J Comput Phys 378:686–707
5. Lumey JL (2012) Stochastic tools in turbulence. Elsevier Science, Burlington
6. Wang S, Yu X, Perdikaris P (2022) When and why PINNs fail to train: A neural tangent kernel perspective. J Comput Phys 449:110768
7. EW, Yu B (2018) The deep Ritz method: A deep learning-based numerical algorithm for solving variational problems. Commun Math Stat 6(1):1–12
8. Cen J, Zou Q (2024) Deep finite volume method for partial differential equations. J Comput Phys 517:113307
9. Sirignano J, Spiliopoulos K (2018) DGM: A deep learning algorithm for solving partial differential equations. J Comput Phys 375:1339–1364
10. Jingrun Chen JC, Xurong Chi XC, Weinan EWE et al (2022) Bridging traditional and machine learning-based algorithms for solving PDEs: The random feature method. J Mach Learn 1(3):268–298
11. Kim J, Lee K, Lee D et al (2021) DPM: A novel training method for physics-informed neural networks in extrapolation. In: Proceedings of the AAAI conference on artificial intelligence, vol 35(9), pp 8146–8154
12. Wang S, Teng Y, Perdikaris P (2021) Understanding and mitigating gradient flow pathologies in physics-informed neural networks. SIAM J Sci Comput 43(5):A3055–A3081
13. Shin Y, Darbon J, Karniadakis GE (2020) On the convergence of physics informed neural networks for linear second-order elliptic and parabolic type PDEs. Commun Comput Phys 28(5):2042–2074
14. McClenny LD, Braga-Neto UM (2023) Self-adaptive physics-informed neural networks. J Comput Phys 474:111722
15. Liu D, Wang Y (2021) A dual-dimer method for training physics-constrained neural networks with minimax architecture. Neural Netw 136:112–125
16. Yu J, Lu L, Meng X et al (2022) Gradient-enhanced physics-informed neural networks for forward and inverse PDE problems. Comput Methods Appl Mech Eng 393:114823
17. Jagtap AD, Kawaguchi K, Karniadakis GE (2020) Adaptive activation functions accelerate convergence in deep and physics-informed neural networks. J Comput Phys 404:109136
18. Sitzmann V, Martel J, Bergman A et al (2020) Implicit neural representations with periodic activation functions. Advances in neural information processing systems, vol 33. Curran Associates, Inc., pp 7462–7473
19. Wang Y, Sun J, Bai J et al (2025) Kolmogorov–Arnold-Informed neural network: A physics-informed deep learning framework for solving forward and inverse problems based on Kolmogorov-Arnold networks. Comput Methods Appl Mech Eng 433:117518

A Method for Predicting the Virtual Energy Storage State of Urban Electric Vehicles Based on a Stochastic Utility Model

Qifang Chen, Hongye Tian, Mingchao Xia, Qianhao Sun, and Yubin Wang

Abstract As electric vehicles (EVs) are increasingly integrated into the power grid, their uncoordinated charging behavior poses challenges to grid stability. Simultaneously, their mobile energy storage characteristics offer new avenues for coordinating vehicle-grid interaction. This paper proposes a method for predicting the virtual energy storage state of urban electric vehicles, establishing a comprehensive framework from travel chain modeling to energy storage state assessment. Steady-state and non-steady-state storage energy, as well as external charging/discharging power, are defined to quantitatively analyze the dynamic storage characteristics of electric vehicle clusters. Simulation results show that the proposed model can not only describe EV travel behavior but also reliably predict the spatiotemporal distribution of charging load at various nodes in the network and assess their potential for energy storage dispatch.

Keywords Electric vehicles · Stochastic utility · Load forecasting · Virtual energy storage

1 Introduction

Driven by the carbon peaking and carbon neutrality goals and energy transition goals, electric vehicles have experienced rapid development and widespread adoption. According to data from the Ministry of Public Security, China now has 22.09 million EVs and 11.884 million charging units [1]. This has created a new ecosystem where EVs, charging points, and the power grid work together. This supports the growth of China's new energy industry. However, the large-scale use of EVs also brings two

Q. Chen · H. Tian · M. Xia (✉) · Q. Sun · Y. Wang
School of Electrical Engineering, Beijing Jiaotong University, Beijing, China
e-mail: mchxia@bjtu.edu.cn

M. Xia
School of Electrical Engineering, Shenyang University of Technology, Shenyang, China

G. Mu et al. (eds.), *Proceedings of the 2nd Electrical Artificial Intelligence Conference, Volume 1*, Lecture Notes in Electrical Engineering 1587,
https://doi.org/10.1007/978-981-95-7932-7_15

main challenges. On one hand, the random nature of user charging behavior poses big threats to grid stability, such as reduced reactive power [2] and increased operational losses [3]. On the other hand, the flexibility of EVs in time and space, along with their mobile energy storage feature, offers a new way to help manage the grid. Groups of EVs can act as flexible units that release stored power when renewable energy output is low and absorb extra power during peak times. This improves the grid's ability to handle renewable energy and manage peak demand. Therefore, accurately characterizing the virtual energy storage state of EVs is a core step for transforming them from potential loads into dispatchable assets.

Regarding EV charging demand forecasting, existing research primarily falls into two categories. The first approach focuses on predicting demand based on behavioral characteristics such as user charging habits, travel schedules, and travel frequency [4–6]. This approach focuses more on actual EV usage behavior but overlooks the fact that, as traffic participants, their journeys are also influenced by traffic conditions. By not considering traffic flow characteristics, the model's output deviates from the actual spatiotemporal movement of EVs. Although reference [7] considers the impact of traffic conditions on travel speed and energy consumption, it assumes trips take the shortest path. It does not fully account for the diversity of user route choices in congested situations, which reduces the accuracy of load forecasting. The second approach extracts charging behavior features from historical load data [8–10], such as charging time, location, and weather, to predict EV charging demand during specific time periods. However, this method focuses solely on local area loads and relies on data from single regions or charging stations, failing to reflect the spatiotemporal distribution differences of EVs across different areas. Consequently, it struggles to s300upport grid-wide dispatch decisions. Furthermore, existing research lacks integrated node-based models combining transportation networks with power grids. They analyze grid node loads or traffic node flows in isolation. This makes it impossible to establish the relationship between traffic flow, power load, and energy storage, and thus prevents accurately measuring the charging and discharging potential of EVs at each node.

To address these shortcomings, this paper constructs an EV virtual energy storage model integrating urban transportation and power systems, enabling real-time and accurate assessment of the regulation potential at each node across the city.

2 Spatiotemporal Distribution Model for EV Charging

Accurately describing user travel behavior is key to forecasting load and studying changes in EV energy storage state. This paper uses trip chain theory to represent the dynamic changes in user travel across both time and space.

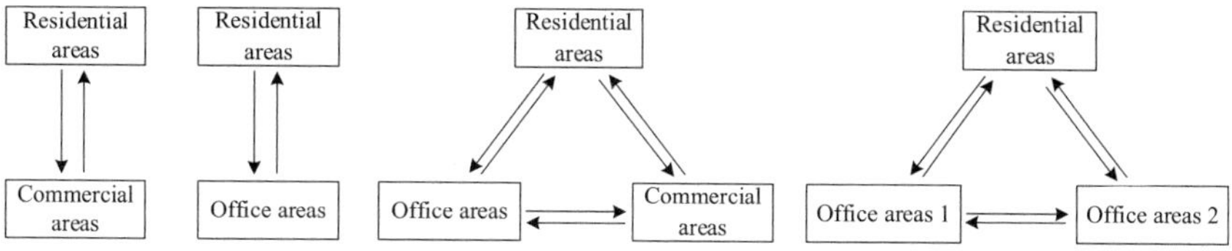

Fig. 1 Diagram of the four trip chains

2.1 Travel Characteristics

Based on the needs of user travel activities, this paper focuses on three types of typical functional areas: residential, office, and commercial areas. The residential area is set as the starting point of the trip chain, and users return to it after completing their daily travel. Considering only private cars as the mode of travel, four typical closed-loop trip chains are constructed, as shown in Fig. 1. The probabilities for these four trip chains are assumed to be 5%, 15%, 50%, and 30% respectively [11]. For the third travel chain, electric vehicles do not strictly follow the fixed path of Residential -Office -Commercial -Residential during transfers. The key point is that both the start and end locations are residential areas. The specific daytime travel itinerary, however, dynamically determines the next destination type based on a transition probability matrix. This means complex trip chains such as Residential-Office-Commercial-Office-Commercial-Residential may occur.

Analysis of the 2017 National Household Travel Survey (NHTS) data shows that the start time of the first trip, the end time of the last trip, and the parking time in office areas for private cars all follow a mixed Gaussian distribution [12]. Meanwhile, the parking time in residential and commercial areas follow a lognormal distribution. For the transition matrix needed in the trip chain model, user travel times are divided into 30-min intervals. By analyzing the distribution of destination types corresponding to EV travel times, the travel transition probability matrix for different time periods can be obtained [13].

2.2 Travel Decision

During EV travel, travel time is affected by the road network and traffic conditions. Therefore, modeling the traffic network must be considered when simulating trip chains. This paper uses graph theory to convert a real road network into nodes and edges. Here, the nodes represent road intersections and their surrounding areas. These nodes are then classified into three types of zones based on their attributes: office, residential, and commercial areas. The resulting road network topology is shown in Fig. 2. In this figure, all edges represent two-way roads. The nodes are color-coded: red for commercial areas, green for office areas, and blue for residential areas.

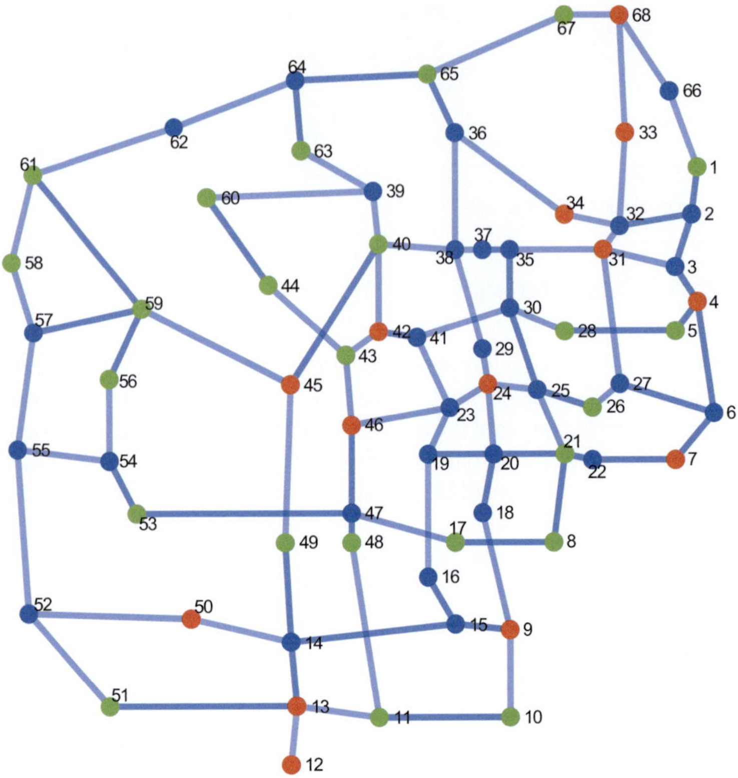

Fig. 2 Diagram of the road network topology

Based on the latitude and longitude coordinates of the nodes, the straight-line distance between nodes (p, q) can be calculated using Eq. (1), where (λ_p, φ_p) and (λ_q, φ_q) represent the latitude and longitude coordinates of nodes p and q respectively. The value 111 denotes that each degree of latitude corresponds to a meridian arc length of 111 km. γ is the curvature coefficient, which accounts for the actual slope of roads and is set to 1.15 [7].

$$l_{pq} = \gamma \times 111 \times \sqrt{\left(\varphi_p - \varphi_q\right)^2 + (\lambda_p - \lambda_q)^2 \cos^2\left(\frac{\varphi_p + \varphi_q}{2}\right)} \tag{1}$$

In real-world road networks, travel time is the primary factor users consider when selecting routes. The U.S. Federal Highway Administration's BPR model is used to reflect the relationship between route travel time and current traffic flow, as shown in Eq. (2).

$$\Delta t_{(p,q)}^{d} = t_{(p,q)}^{free} \times \left(1 + \alpha_{(p,q)} \times \left(\frac{flow_{(p,q)}^{t-1}}{cap_{(p,q)}}\right)^{\beta_{(p,q)}}\right) \tag{2}$$

where $\Delta t_{(p,q)}^{d}$ represents the travel time on road (p, q), $t_{(p,q)}^{free}$ denotes the free travel time on road (p, q), $flow_{(p,q)}^{t-1}$ indicates the traffic flow on road (p, q) at the previous time step, $cap_{(p,q)}$ signifies the capacity of road (p, q), and $\alpha_{(p,q)}$ and $\beta_{(p,q)}$ are congestion coefficients typically set to 0.15 and 4, respectively.

When choosing a route, users also consider factors such as distance and traffic congestion to minimize their travel cost. However, due to individual differences and incomplete traffic information, users do not always choose the path with the shortest travel time. Therefore, this paper uses the *Logit* random utility model to describe user route choice behavior, as shown in Eqs. (3) and (4).

$$p_{k,t}^{rs} = \frac{\exp(-\theta C_{k,t}^{rs})}{\sum_{k} \exp(-\theta C_{k,t}^{rs})}, k \in K_{rs} \tag{3}$$

$$C_{k,t}^{rs} = \sum_{(p,q)} \Delta t_{(p,q)}^{d} \cdot \varepsilon_{pq,k}^{rs}, \varepsilon_{pq,k}^{rs} = \begin{cases} 0, (p, q) \notin k \\ 1, (p, q) \in k \end{cases} \tag{4}$$

where K_{rs} denotes the feasible path set between origin node r and destination node s, $p_{k,t}^{rs}$ represents the probability of selecting path k, $C_{k,t}^{rs}$ is thetravel cost of path k, $\varepsilon_{pq,k}^{rs}$ is the association coefficient between path k and segment (p, q), and θ indicates the user's awareness of current traffic conditions. For commuting-centric trips between residential and office nodes, where users exhibit strong daily regularity and high familiarity with the transportation network $\theta = 0.8$ [14], whereas for trips involving commercial area nodes, considering users' limited awareness of traffic conditions on non-routine routes and greater random influence on path selection, θ is set to 0.6.

After determining the starting node and destination node, Yen's Algorithm is used to find the set of routes between them. This algorithm can generate the top K shortest valid paths in order of length. Its method of building restricted sets ensures that each new path includes at least one road segment not used by any previous path. This prevents the generation of duplicate routes and provides a set of distinct candidate paths with practical differences for the Logit random utility model.

2.3 Charging Method Selection and Decision

The distribution of electric vehicle charging demand and magnitude is influenced by multiple factors including vehicle parameters, user travel patterns, and charging habits. The preceding travel chain model accurately captures the temporal and spatial distribution characteristics of vehicles. By tracking vehicle trajectories and stationary locations across different time periods, it clearly reveals when and where charging demand arises. Based on the spatiotemporal distribution of vehicle charging demand, this paper proposes a user charging decision scheme, laying the foundation for establishing an unordered charging model for electric vehicles.

To simplify model calculations, this paper categorizes EV charging methods into fast charging and conventional charging, both modeled as constant-power charging. Having obtained the specific status and location of EVs at each time point, the charging mode selected by the kth EV to meet the ith trip demand is determined by the entry charge level $SOC_{k,i-1}^{a}$, parking time $\Delta t_{k,i-1}^{p}$, and required exit charge level $SOC_{k,i}^{need}$, as calculated by Eqs. (5) and (6). The entry charge level and parking time are determined by the $i-1$th trip.

$$SOC_{k,i}^{need} = SOC_{k,i}^{miss} + SOC_{k}^{\min} \tag{5}$$

$$SOC_{k,i}^{miss} = (P_{k}^{c} \times dis_{k,(i-1,i)})/Q_{k}^{bat} \tag{6}$$

where $SOC_{k,i}^{miss}$ represents the charge consumed during the ith trip, $SOC_{k}^{\min}$ denotes the minimum charge level, P_{k}^{c} indicates the distance of the ith trip, Q_{k}^{bat} signifies the energy consumption per kilometer for the kth vehicle, and Q_{k}^{bat} represents the battery capacity of the kth vehicle.

Given the required charge for departure, first calculate the charging duration at the standard charging power P_{k}^{gen}:

$$t_{i}^{ch,gen} = \begin{cases} 60 \times (SOC_{k,i}^{need} - SOC_{k,i-1}^{a})Q_{k}^{bat}/P_{k}^{gen}, SOC_{k,i}^{need} > SOC_{k,i-1}^{a} \\ 0, SOC_{k,i}^{need} \leq SOC_{k,i-1}^{a} \end{cases} \tag{7}$$

Equation (7) indicates that if the entry charge level $SOC_{k,i-1}^{a}$ meets the charge level required for the next trip $SOC_{k,i}^{need}$, no charging occurs.

Next, evaluate the relationship between the required charging time $t_{i}^{ch,gen}$ and the available parking time $\Delta t_{k,i}^{p}$. If $t_{i}^{ch,gen} \geq \Delta t_{k,i}^{p}$, select standard charging; if $t_{i}^{ch,gen} < \Delta t_{k,i}^{p}$, select fast charging. Once charging commences, set a minimum target of $SOC_{k,i}^{need}$ or cease charging upon reaching the maximum charge level $SOC_{k}^{\max}$.

3 Virtual Energy Storage Model

The core idea of electric vehicle virtual energy storage is to treat a large number of scattered EVs as a single, unified virtual storage unit. By using smart control technology to coordinate their charging and discharging behavior, it enables the centralized management and efficient use of distributed power resources. This paper considers two key storage parameters: energy capacity and charging/discharging power. It closely combines the trip chain model with the EV charging/discharging process to establish a virtual energy storage model for EVs.

3.1 Virtual Energy Storage Capacity

(1) Steady-State Energy

Steady-state energy is defined as the sum of the products of the current SOC and total battery capacity for all vehicles currently parked at a node. The steady-state value of virtual energy storage at node n during the time interval $[t_0, t_0 + T^{disp}]$ is calculated as Eq. (8).

Based on the travel chain, determine the vehicle's location at each time point. Identify the number of vehicles $num_n^{storage}$ that remain stationary at node n throughout the time interval. Using the electric vehicle charging demand model, obtain the average SOC value for the vehicle during the corresponding time interval. Multiply this value by the vehicle's battery capacity to calculate its energy storage capacity for that time interval.

$$Q_n^{BAT_w} = \sum_{j=1}^{num_n^{storage}} \left(\sum_{t=t_0}^{t_0+T^{disp}} SOC_j^t \right) / T^{disp} * Q_j^{bat} \tag{8}$$

where Q_j^{bat} represents the battery capacity of the electric vehicle labeled j within the cluster of parked vehicles at node n, and T^{disp} (min) is defined as the virtual energy storage scheduling cycle for the cluster of electric vehicles.

(2) Non-Steady-State Energy

This paper defines Non-Steady-State energy as the total energy change resulting from vehicles entering or leaving a charging node. Based on vehicle location and state of charge, during the time interval $[t_0, t_0 + T^{disp}]$, if any electric vehicle newly enters the jurisdiction of node n but has not yet connected to node n for charging, or has completed charging but remains within the jurisdiction of node n, then the value of node n's Non-Steady-State energy storage $Q_n^{BAT_z}$ is:

$$Q_n^{BAT_z} = \sum_{h=1}^{num_n^{access}} SOC_h^{t_h} * Q_h^{bat} - \sum_{d=1}^{num_n^{leave}} SOC_d^{t_d} * Q_d^{bat} \tag{9}$$

where $t_h,\ t_d \in (t_0, t_0 + T^{disp})$ and num^{access} denote the number of vehicles charging at node n during the time interval, $SOC_h^{t_h}$ represents the SOC of vehicle h at its connection time t_h, and Q_h^{bat} is the battery capacity of electric vehicle h within the vehicle cluster at node n. num^{leave} indicates the number of vehicles leaving node n during the time interval, $SOC_h^{t_d}$ denotes the SOC of vehicle d at its departure time

t_d, and Q_d^{bat} is the battery capacity of electric vehicle d within the vehicle cluster at node n.

Then, the virtual energy storage capacity Q_n^{BAT} at node n at time t_0 is:

$$Q_n^{BAT} = Q_n^{BAT_w} + Q_n^{BAT_z} \tag{10}$$

3.2 Virtual Energy Storage Charge/Discharge Power

The external charge/discharge power of an electric vehicle cluster is constrained by the current SOC of individual vehicles and their next trip schedules. Essentially, the cluster's external charge/discharge power represents the aggregate charging/discharging capacity of all individual vehicles. This paper first determines the external charge/discharge capacity of each individual vehicle. By summing these capacities, the total cluster capacity is obtained. This total capacity is then divided by the dispatch cycle duration to derive the cluster's external charge/discharge power.

The method for determining the maximum charge capacity of an individual electric vehicle is as follows:

$$Q_k^{ch_1} = P_k^{\max} T^{disp}/60 \tag{11}$$

$$Q_k^{ch_2} = (SOC_k^{\max} - SOC_k^{now})Q_k^{bat} \tag{12}$$

$$Q_k^{ch_3} = P_k^{\max}(t_{i,k}^s - t_{now})/60 \tag{13}$$

where $Q_k^{ch_1}$ represents the maximum chargeable capacity of vehicle k under the scheduling cycle constraint T^{disp}; $Q_k^{ch_2}$ denotes the maximum chargeable capacity of vehicle k under the maximum state-of-charge standard constraint $SOC_k^{\max}$; $Q_k^{ch_3}$ indicates the maximum chargeable capacity under the planned departure time constraint $t_{i,k}^s$.

The maximum chargeable capacity Q_k^{ch} of vehicle k at time t_{now} is determined by taking the minimum value among the three constraints, as shown in Eq. (14).

$$Q_k^{ch} = \min(Q_k^{ch_1}, Q_k^{ch_2}, Q_k^{ch_3}) \tag{14}$$

The determination method for the lower bound of discharge capacity for a single electric vehicle is as follows:

$$Q_k^{dch_1} = -P_k^{\max} T^{disp}/60 \tag{15}$$

$$Q_k^{dch_2} = -(SOC_k^{now} - SOC_k^{\min})Q_k^{bat} \tag{16}$$

$$Q_k^{dch_3} = -P_k^{\max}(t_{i,k}^s - t_{now})/60 \tag{17}$$

$$Q_k^{dch_4} = -\frac{Q_k^{bat}}{2}[SOC_k^{now} - SOC_{i,k}^{need} + \frac{P_k^{\max}}{60Q_k^{bat}}(t_{i,k}^s - t_{now})] \tag{18}$$

where $Q_k^{dch_1}$ represents the maximum discharge capacity of vehicle k under the scheduling cycle constraint T^{disp}; $Q_k^{dch_2}$ denotes the maximum discharge capacity of vehicle k under the minimum charge standard constraint $SOC_k^{\min}$; $Q_k^{dch_3}$ indicates the maximum discharge capacity under the planned departure time constraint $t_{i,k}^s$; $Q_k^{dch_4}$ signifies the maximum discharge capacity under the next trip requirement constraint $SOC_{i,k}^{need}$.

Then, the maximum discharge capacity Q_k^{dch} of vehicle k at time t_{now} is given by Eq. (19).

$$Q_m^{dch} = \max(Q_m^{dch_1}, Q_m^{dch_2}, Q_m^{dch_3}, Q_m^{dch_4}) \tag{19}$$

Based on the travel chain $[t_0, t_0 + T^{disp}]$, determine the number of electric vehicles $num_n^{storage}$ parked at node n during the time period. By accumulating the external charge/discharge capacity of each electric vehicle, the steady-state values of the cluster electric vehicle virtual energy storage charge capacity $Q_n^{CH_w}$ and discharge capacity $Q_n^{DCH_w}$ at node n during the time period can be obtained.

$$Q_n^{CH_w} = \sum_{m=1}^{num_n^{storage}} Q_m^{ch} \tag{20}$$

$$Q_n^{DCH_w} = \sum_{m=1}^{num_n^{storage}} Q_m^{dch} \tag{21}$$

If during the time period $[t_0, t_0 + T^{disp}]$, an electric vehicle enters the jurisdiction of node n but has not yet connected to node n, and either begins charging or has completed charging but remains within the jurisdiction of node n, then the non-steady-state values for node n's chargeable capacity $Q_n^{CH_z}$ and dischargeable capacity $Q_n^{DCH_z}$ are:

$$Q_n^{CH_z} = \sum_{h=1}^{num_n^{access}} Q_h^{ch} - \sum_{d=1}^{num_n^{leave}} Q_d^{ch} \tag{22}$$

$$Q_n^{DCH_z} = \sum_{h=1}^{num_n^{access}} Q_h^{dch} - \sum_{d=1}^{num_n^{leave}} Q_d^{dch} \tag{23}$$

Then, during the time period $[t_0, t_0 + T^{disp}]$, the chargeable capacity Q_n^{CH}, dischargeable capacity Q_n^{DCH}, external charging power P_n^{CH}, and external discharging power P_n^{DCH} of node n are:

$$Q_n^{CH} = Q_n^{CH_w} + Q_n^{CH_z} \tag{24}$$

$$Q_n^{DCH} = Q_n^{DCH_w} + Q_n^{DCH_z} \tag{25}$$

$$P_n^{CH} = Q_n^{CH} / T^{disp} \tag{26}$$

$$P_n^{DCH} = Q_n^{DCH} / T^{disp} \tag{27}$$

4 Simulation Results and Analysis

4.1 Simulation Parameters

The simulation input included a 68-node road network topology map comprising 14 commercial area nodes, 24 office area nodes, and 30 residential area nodes. The simulation parameters were set as follows: $N = 3000$, $T = 10,080$, $K = 5$, $SOC^{\min} = 0.2$, $SOC^{\max} = 0.9$, $T^{disp} = 5$. To assess the impact of different electric vehicle models on the model, parameters for various vehicle types are detailed in Table 1.

Table 1 Basic parameters of electric vehicles

	BYD Qin Pro	BAIC EU5	Xpeng G3
Battery capacity (kWh)	56.4	61.8	57.5
Fast charging power (kw)	35	60	90
Standard charging power (kw)	7	7	7
Energy consumption per kilometer (kWh/km)	0.205	0.136	0.155

4.2 Simulation Results

Results from Days 3–5 were analyzed. The variations in load and virtual energy storage state at each node are shown in the figure below.

Regarding the node load variations shown in Fig. 4, aggregating node loads by their respective zones reveals three distinct patterns: Residential zones exhibit load peaks between 11:00–12:00 and 20:00–21:00; Office zones peaked at 9:00 and 13:00–14:00, primarily due to increased user arrival for work activities driving corresponding load growth; Commercial zones peaked between 11:00–12:00 and 20:00, mainly reflecting post-work or end-of-day activities. From a node perspective, loads between 6:00 and 24:00 are concentrated at nodes 9–13, 31–35, and 49–51. Nodes 9–13 and 49–51 are located within the commercial-office zone, indicating that this area experiences frequent commercial and office activities, high pedestrian and vehicle traffic, and significant electric vehicle charging demand. Nodes 31–35 belong to residential-commercial zones. Vehicles in residential areas require charging after completing daily trips, while commercial activities also generate charging demand. The combined effect of these two factors leads to concentrated load at these nodes. From a temporal perspective, compared to other periods, all nodes exhibit lower charging load values between 2:20 AM and 6:00 AM. This indicates that most vehicles returning to residential nodes to begin charging have completed their charging processes by this time (Fig. 3).

Analyzing the results in Fig. 4, the steady-state virtual storage energy values at each node visually show alternating red and blue patterns over time, with nodes of the same type changing synchronously. In terms of the virtual storage energy distribution,

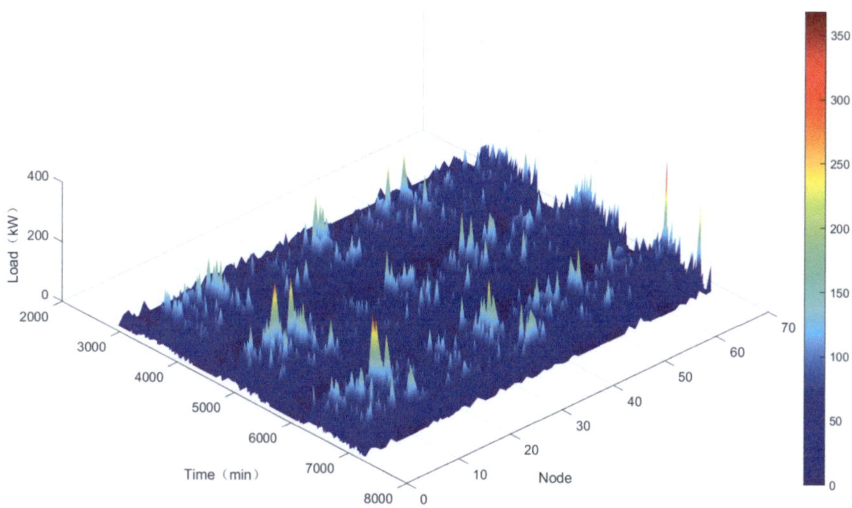

Fig. 3 Load distribution for each node (Days 3–5)

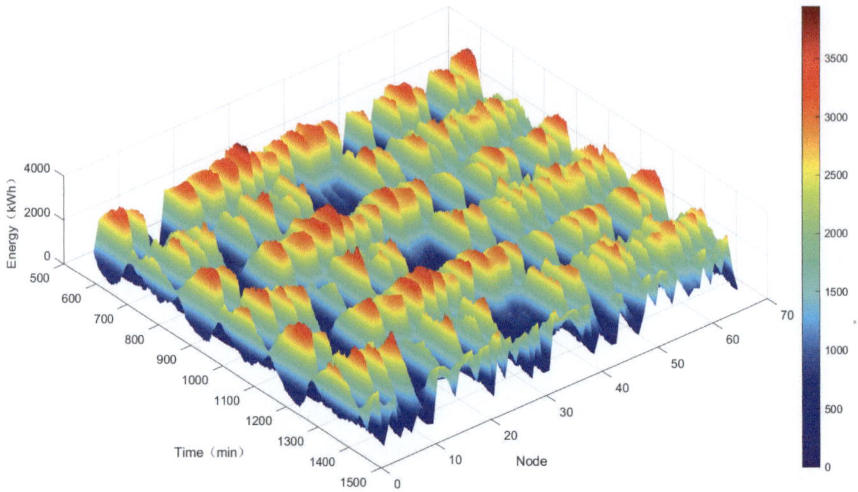

Fig. 4 Steady-State virtual energy storage distribution for each node (Days 3–5)

residential nodes and office nodes exhibit a clear complementary pattern: when residential nodes reach their peak energy, office nodes are at their lowest, and vice versa. In contrast, commercial nodes generally maintain lower energy levels with minimal fluctuation. This indicates that virtual storage energy is primarily concentrated in residential and office areas. This aligns with real-world conditions, as vehicles typically remain parked for extended periods in these zones during weekdays.

The values in Fig. 5 are mostly between -100 and 100 kWh, indicating low vehicle mobility in this road network. Analyzing by scheduling periods, during periods 576–637, 863–929, and 1146–1215, the non-steady-state virtual energy storage values at each node show clear lows. This indicates fewer vehicles are traveling during 23:30–5:15 on weekdays, which matches the real-world situation.

Figure 6 shows that nodes in residential areas have high available discharging power to the grid at the start of the day. This is because most vehicles are parked and have not yet started their daily travel. From around 7:00, as residents begin their trips, this available power drops rapidly. It remains at a low level until about 22:30, after which it starts to increase again, reaching a maximum of 7300 kW. Office area nodes show the opposite pattern. Their available discharging power is concentrated during the daytime, creating a complementary peak-and-valley pattern with the residential areas. In contrast, the available discharging power for commercial area nodes remains below 1500 kW. It stays at this low level with little variation. These patterns indicate that on weekdays, resident travel primarily involves commuting between home and the office. Fewer vehicles park in commercial areas, and their stay duration is shorter, resulting in lower available discharging power.

Figure 7 shows that the change trend of the discharge power available to the grid at each node is basically the same as that of the charging power. It also shows the complementary peak-and-valley pattern between residential and office nodes, while

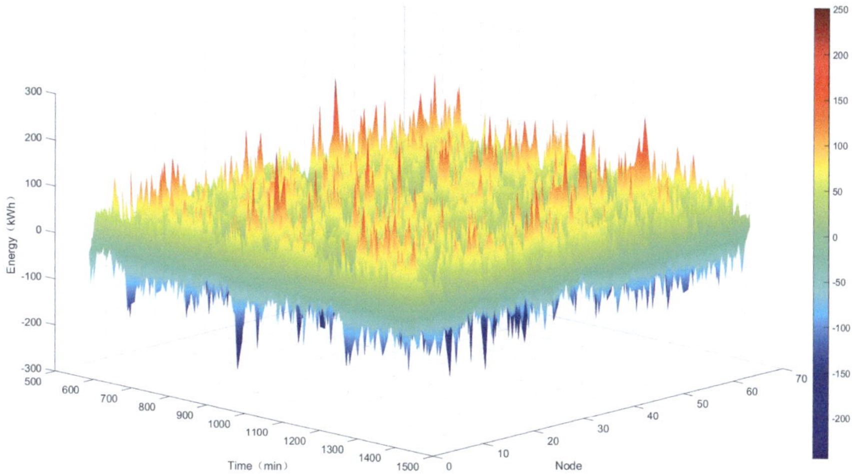

Fig. 5 Non-steady-state virtual energy storage distribution for each node (Days 3–5)

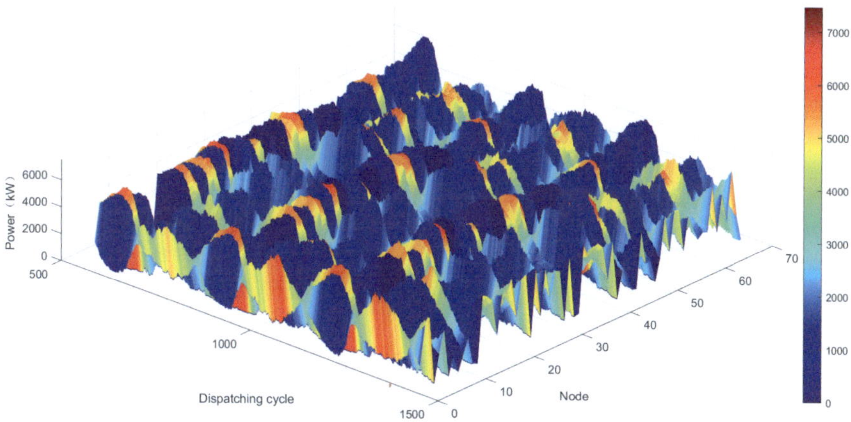

Fig. 6 Charging power available to the grid for each node (Days 3–5)

commercial nodes maintain low values with little variation. The difference is that the available discharge power is limited by the energy needed for the vehicles' next trip. Therefore, the discharge power values are generally lower than the charging power values. Furthermore, as vehicles depart, the peak discharge power in residential nodes decreases. For example, at Node 2, the maximum discharge power can reach 3982.74 kW during scheduling periods 576–660, but it drops to 3321.85 kW during periods 846–948. The discharge power in office nodes during the daytime shows a pattern of first increasing and then decreasing. Taking Node 5 as an example, during scheduling periods 701–826, the discharge power initially increases, peaks at 3061.38 kW by period 773, and then starts to decrease. This indicates that the EV cluster at this node

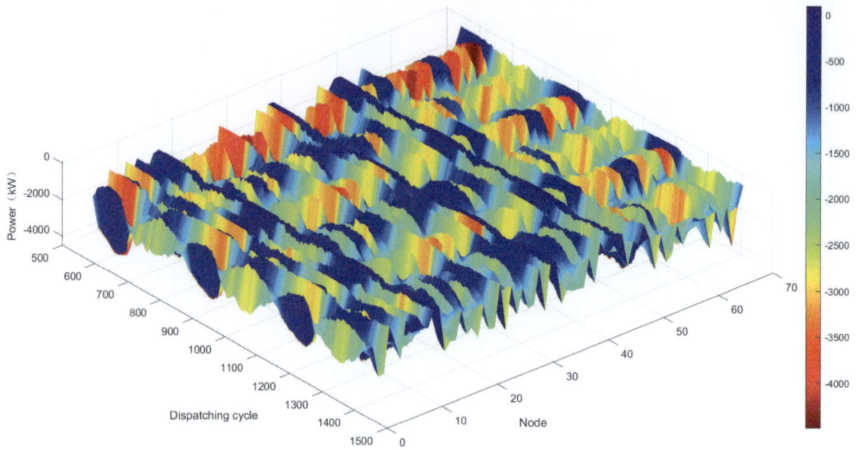

Fig. 7 Discharge power available to the grid for each node (Days 3–5)

has sufficient SOC for discharging after 10:25. After 16:25, vehicles begin to leave this node, causing the available discharge power to drop.

5 Conclusions

This research addresses the problem of predicting the virtual energy storage state of urban electric vehicles by proposing an analysis method based on a random utility model. It systematically describes the dynamic energy storage characteristics of EV clusters within the coupled transportation-power network.

Simulation results show that the spatiotemporal distribution of charging load is highly coupled with user travel behavior. Load peaks in residential, office, and commercial area nodes all correspond to their typical activity patterns, verifying that the trip chain model effectively captures charging demand. Regarding storage energy, nodes show periodic fluctuations. A significant spatiotemporal complementary characteristic exists between residential and office nodes, reflecting the impact of the home-work commute pattern on energy storage distribution. In terms of charging and discharging capability, different functional areas show complementarity, while nodes of the same type show synergy. Discharge power, constrained by the energy required for the next trip, is generally lower than charging power. The model can accurately predict the spatiotemporal distribution of charging load and the potential for energy storage dispatch at road network nodes, providing theoretical support for precise grid management and dispatch.

Acknowledgements This work is supported by the National Natural Science Foundation of China [No. 52277074].

References

1. Central People's Government of the People's Republic of China Website. https://www.gov.cn/zhengce/jiedu/tujie/202501/content_6999527.htm. Last accessed 12 May 2025
2. Li H, Huang X, Chen Z et al (2015) Reliability assessment of distribution networks incorporating wireless charging for electric vehicles. Trans Chin Soc Electr Eng 30(S1):244–250
3. Wang F, Li Z, Ye M et al (2016) Review of the impact of electric vehicle charging on power grids and research on optimal scheduling. South Power Grid Technol 10(06):70–80
4. Li F, Dou C, Xu S (2019) Optimal scheduling strategy of distribution network based on electric vehicle forecasting. Electronics 8(7):816
5. Yi T, Zhang C, Lin T et al (2020) Research on the spatio-temporal distribution of electric vehicle charging load demand: A case study in China. J Clean Prod 24(21):118457
6. Zhao Y, Wang Z, Shen Z et al (2021) Data-driven framework for large-scale prediction of charging energy in electric vehicles. Appl Energy 282:116175
7. Song Y, Lin S, Tang Z et al (2020) Probabilistic modeling of spatiotemporal distribution of electric vehicle charging load based on dynamic traffic flow. Autom Electr Power Syst 44(23):47–56
8. Huang N, Liu D, Cai G et al (2021) Interval prediction of electric vehicle charging load based on multi-correlated day scenario generation. Trans Chin Soc Electr Eng 41(23):7980–7990
9. Wang Y, Gu Y, Ding Z et al (2020) Electric vehicle charging demand forecasting based on fuzzy entropy and ensemble learning. Autom Electr Power Syst 44(03):114–121
10. Zhu J, Yang Z, Guo Y et al (2019) Short-term load forecasting for electric vehicle charging stations based on deep learning approaches. Appl Sci 9(9):1723
11. Zhang M, Sun Q, Yang X (2022) Electric vehicle charging load forecasting considering real-time interaction of multi-source information and user regret psychology. Power Syst Technol 46(02):632–645
12. Aghajan-Eshkevari S, Ameli MT, Azad S (2023) Optimal routing and power management of electric vehicles in coupled power distribution and transportation systems. Appl Energy 341:121126
13. Wang T (2022) Modeling and cooperative optimization control methods for electric vehicle virtual energy storage. Beijing Jiaotong University, pp 17–20
14. Ling C (2017) Research on an improved logit-type random user balanced allocation model and algorithm. Lanzhou Jiaotong University, pp 31–34

Smart Operation of EV Clusters with Active Distribution Network Based on LLM-MBPO Cooperation

Yang Zhou, Ran Wei, Yinan Hong, Liang Sun, and Hao Xu

Abstract With fast development and penetration of Electric Vehicles (EVs), the charging demand of EVs has grow over tenfold in recent years in major citys. In this work, we aim at the issues raised by the EV clusters charging demand, and proposed a LLM-MBPO Cooperation framework for smart operations of EV clusters with active distribution grids. First, a large language model (LLM) with model-based policy Optimization (MBPO) framework is developed. The comprehensive interactive capability of LLM is used to bridge the gap between the uncertain charging operations and the model settings of MBPO. Second, the LLM is fine tuned power distribution system operation and optimization knowledge, to fit it the role of regulating the MBPO. The MBPO is continuous action space deep reinforcement learning algorithm, which is managing the charging operations of EV clusters. Lastly, the proposed framework is tested with case studies and verified that it can effectively save the charging cost of EV clusters while providing operation supports to the distribution grid.

Keywords Electric vehicle · Active distribution grid · Smart grid · Large language model · Deep reinforcement learning

1 Introduction

Under the "dual carbon goals" of China, the electric vehicle (EV) industry is developing at an astonishing speed. People lived in major cities are pursuing new models of EVs and expecting longer ranges and faster charging experiences. With the new EV models from major manufactures, the power speed of a single EV can reach 1 MW with 800v DC. And the state-of-the-art chargers can provide over 1.5 MW charging speed. Even though this can improve the users experience, it creates significant challenges for the operations of distribution systems. The first challenge is high volume of power flow, which may over the capacity or operation limit of distribution grid.

Y. Zhou · R. Wei · Y. Hong (✉) · L. Sun · H. Xu
State Grid Beijing Changping Power Supply Company, Beijing, China
e-mail: infi@sina.com

G. Mu et al. (eds.), *Proceedings of the 2nd Electrical Artificial Intelligence Conference, Volume 1*, Lecture Notes in Electrical Engineering 1587,
https://doi.org/10.1007/978-981-95-7932-7_16

Second challenge is the voltage issue, as their high-power, short-duration charging cycles can cause rapid voltage fluctuations and sags, potentially destabilizing the local power network. Third challenge is the uncertainty of the charging demand. Together with distributed energy resources, the uncertainties are echoing in the distribution system. These lead to severe operation challenges in active distribution systems (ADNs).

1.1 Literature Review

To address these issues, researchers and engineers from academia and industry have proposed many comparative solutions and mathematical models. Literature related can be classified into two main categories.

The first category includes formulating the problem as a conventional optimization problem, convex or non-convex optimizations, and solving them with customized developed algorithms, commercial solvers, or heuristic algorithm. A virtual power plant with large scale wind power and EV fleet is formed, and the optimization configuration and approaches includes wavelet packet decomposition and particle swarm algorithm [1]. The V2G potential of EV groups are integrated into power system frequency regulations in [2]. It also provide suggestions on the development of charging stations and EV charging/discharging plans. In terms of coordinated dispatch of power and transportation system, Ref. [3] developed an interesting framework to achieve the maximal social welfare of the two networks, by considering the hydrogen energy storage system. An improve heterogeneous decomposition algorithm is used to solve the optimization problem. The impact of EVs on power quality, voltage profile, and frequency stability of the power system is investigated in [4], the capability of EV to support the system frequency stability is also studied. Researches from Tsinghua university and Imperial College of London introduced a new carbon emission flow model to study the carbon-aware EV power scheduling problem [5]. The case studies showed the evidence for realising the UK's decarbonisation ambitions by 2030 and 2050. In [6], a hierarchical transactive energy framework is develop to locally coordinate EV charging demand and the distributed generation resources (DER) in distribution grid. The problem formulation in [6] is close to this work, and a modified version of the alternating direction method of multipliers is used to solve the problem. The power system resilience is also studied as a optimization problem with the considerations of EV charging/discharging operations and other resources [7], standardized solvers are used to find the optimal solutions. Sum up the above mentioned works, the operation of EV charging has been comprehensively studied with conventional mathematical tools and good results have been achieved in scenarios with less real-time requirement and not much uncertainties.

However, when it comes to the case that efficient operation decisions are needed in real-time with complex, or even non-convex situations and considering large amount of uncertainties, conventional optimization method struggled to achieve satisfied results. In this case, the second category of machine learning-based method played a vital role in recent years. In [8] the co-optimization of power and transportation network with EV is achieved by applying the model-free deep reinforcement learning algorithm. The similar hierarchical EV charging control optimization problem is modeled as a Markov decision process (MDP) and solved by multi-objective reinforcement learning algorithm [9]. With the consideration of enormous dynamics and uncertainties of EV and RES in distribution system, Ref. [10] proposed a multi-agent reinforcement learning approach. In cooperating the graph characteristics of the power and transportation network, multi-agent graph reinforcement learning approach is also deployed to solve the real-time EV requests in [11]. From a perspective of distribution network congestion management considering EVs, Ref. [12] applied constrained MDP and safe reinforcement learning approach. In [13], the authors developed a deep reinforcement learning approach for EVs as prosumers, scheduling and balancing EV autonomy with optimal policies.

1.2 Contributions of This Work

Previous works have paved a good foundation in terms of the operations of EVs and ADNs. The conventional optimization methods can find the optimal solution with well-formulated objective function and constraints. The machine learning-based methods can provide timely result with complex scenarios in real-time. However, the gap is still exist between theoretical research and real world implementation. The charging demand of EV is constantly changing, same as the DERs in the distribution system. These uncertainties and fluctuations create enormous challenge to the state-of-the-art approaches. The generalization ability of machine learning algorithms is limited and the process of fine-tuning requires deep understanding of the developed algorithm, which is hard for the front-line power engineers to modify it on daily basis. It this work, we try to bridge this gap by proposing a large language model and deep reinforcement learning cooperation framework to improve the generalization ability and overcome the challenge of uncertain scenarios. The contributions of this work are the following.

1. To overcome the above generalization issues, a large language model (LLM) and model-based policy optimization (MBPO) cooperation framework is proposed. A per-trained LLM is used to follow the instructions from the system operator and fine-tune the MBPO for online policy update.
2. The power system operation decisions need to be made in timely fashion. In order to improve the policy update speed, model-based policy optimization deep reinforcement learning algorithm is proposed.
3. The proposed LLM-MBPO framework is verified with real world EV charging data and operation scenarios.

The framework is proposed in Sect. 2, Section 3 introduce the LLM, and the MBPO is presented in Sect. 4. Section 5 gives the case studies and Sect. 6 concludes this work.

2 The Large Language Model and Model-Based Policy Optimization Cooperation Framework

In power system operations under scenarios of today's system, the penetrations of solar and wind generations, the implementations of distributed resources and controllable loads and the fast growing of EVs create enormous complexity, huge operation optimization space and multi-source uncertainties into the power system. Conventional optimization formulations may not handle all of the complex details in the modeling. And the AI-based decision making tools are suffered from performance issues in constantly changing environment. Researches have developed sophisticated algorithms to solve these issues. However, the implementation and operations of theses approaches in daily basis require deep knowledge of the specific model and understanding of the engineering problem. This limits the implementation of the deployment of these approaches.

In this work, we proposed the following framework. A LLM is pre-trained with engineering knowledge of the power system operation practice, and it is used as a bridge between the deep reinforcement learning-based solver and operator.

In Fig. 1, ① represents that the LLM is built based on the power system operation knowledge, including the utility-specific dataset, power engineering documents, historical practice and event reports. The LLM can learn the engineering knowledge through this process. ④ and ⑤ represents the classic deep reinforcement learning loop between the policy agent and the training environment. To improve the training process, ⑥ indicates that MBPO agent is using a experience buffer to learn from the engineering experience and the experience from the training process of itself. ② tells that this experience is monitored and also updated by the LLM. ⑨ is the output of the MBPO operation decision to the real world power system. When the scenarios in the real power system changes, the system operation identified this change in ⑩, and then exchange this information with LLM through ③. The LLM gain the information from the system operator, check with the experience buffer, the MBPO agent and the training environment. Consequently, update the reward function of MBPO agent and the training environment to perform a LLM lead transfer learning to improve the generalization capability the decision making agent. This is a novel framework that utilize the capability of LLM and deep reinforcement learning agent to cooperatively adapt to the changing uncertain environment.

3 The Large Language Model and Power System Knowledge Pre-training

The domain-specific knowledge representation of an LLM is stored within high-dimensional weight parameter matrices. Traditional full-parameter fine-tuning methods suffer from low computational efficiency and high resource consumption. In contrast, Low-Rank Adaptation (LoRA) is a fine-tuning method based on

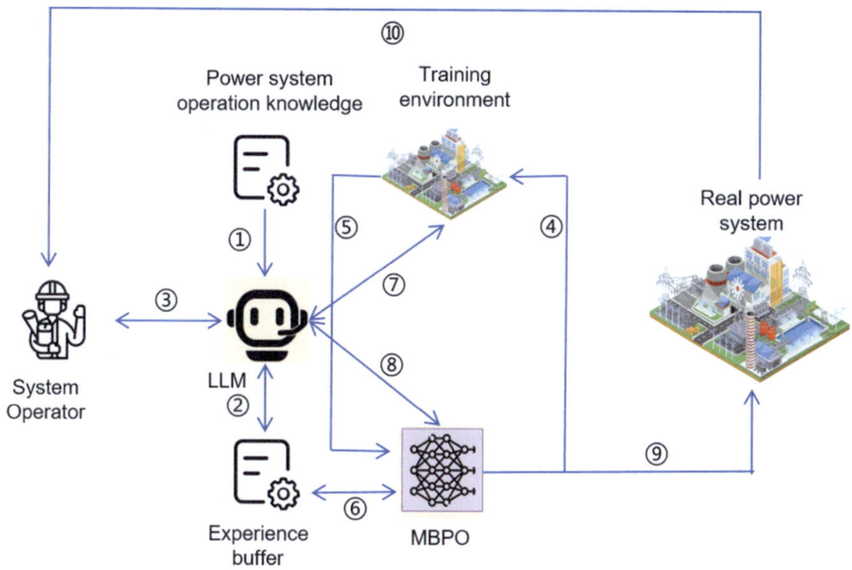

Fig. 1 The proposed LLM-MBPO framework

reparameterization that approximates the performance of full-parameter fine-tuning by simulating the process with low-rank matrices. The information exchange in and out LLM is subsequently standardized into JSON format interface files, which are utilized for secondary adjustments and the development of smaller, specialized models and their operational environments. Conventional LLMs are deep learning language models based on the Transformer architecture, pre-trained on extensive general-purpose text corpora. This enables them to acquire rich semantic information and generate expressions akin to human language. Building upon this foundational general knowledge, this study investigates the fine-tuning of an LLM by incorporating domain-specific corpora from the power system maintenance sector. We employ QLoRA for instruction fine-tuning to derive the LLM.

$$\begin{cases} W_0 + \Delta W = W_0 + BA \\ W_0 \in \mathrm{R}^{d \times k} \\ B \in \mathrm{R}^{d \times r} \\ A \in \mathrm{R}^{r \times d} \\ r << \min(d, k) \end{cases} \tag{1}$$

where W_0 is the pre-trained weight parameter matrix, ΔW is the weight update matrix used to store newly acquired domain-specific knowledge. Therefore, by freezing the base pre-trained weights, LoRA only requires training the two low-rank matrices B and A using power system-specific corpora, such as grid dispatching terminology and balancing meeting experience. This approach enables the effective adaptation of the LLM to the power system maintenance domain while significantly reducing the number of trainable parameters.

$$h = W_0x + \Delta Wx = W_0x + BAx \tag{2}$$

where h is the hidden state tensor, and x is the vectorized representation of the input language from the power maintenance domain. During this propagation process, the LLM, through the newly added weight update matrix BA, can map power system documents to operations of ADN with EVs. It then leverages its original general capabilities to orchestrate the adjustments of smaller models. Building upon LoRA, QLoRA utilizes 4-bit quantization to optimize memory usage [14]. The original pre-trained model is stored in 4-bit precision, while the newly added LoRA weight update matrices are still trained in 16-bit precision. This process optimizes the smaller MBPO model and analyzes the resulting outcomes, ensuring the coherence of operations and improving the generalization ability of the entire framework.

4 The Model-Based Policy Optimization for ADN Operation with EVs

Model-Based Policy Optimization (MBPO) is a reinforcement learning algorithm that strategically bridges the gap between the sample efficiency of model-based methods and the performance stability of model-free approaches. Traditional model-free methods learn policies directly from environmental interactions, requiring vast amounts of data, whereas model-based methods can first learn a ADN operation decision model of the training environment, a process that can be compromised by model bias. MBPO resolves this trade-off by decoupling the roles of model learning and policy optimization. It uses a learned ensemble of probabilistic dynamic operation decision models to generate decision trajectories, which are then used to augment a limited set of real grid operation data. This combined dataset is fed to a high-performance, off-policy model-free algorithm for policy updates. By limiting the policy's exposure to potentially flawed model predictions and leveraging the model primarily for data augmentation, MBPO achieves superior sample efficiency while maintaining the stability and asymptotic performance of leading model-free methods.

The MBPO agent has a internal virtual environment and periodically update the environment model inside with the base off-policy. The MBPO also has an actor and a critics agent. Like other reinforcement learning algorithms. It also has a observation space, an action space and a reward function. The loss function is divide by the number of elements in its corresponding observation specification.

$$Loss = \sum_{i=1}^{N_o} \frac{1}{M_{oi}} Loss_{oi} \tag{3}$$

where N_o and M_o define the parameter of the observation states. The loss is defined as a weighted cross-entropy loss function to deal with the heavily imbalanced data.

$$Loss = \frac{-1}{M}\sum_{i=1}^{M}(w_0 T_i \ln Y_i + w_1(1 - T_i)\ln(1 - Y_i))$$
$$w_0 = \frac{1}{\sum_{i=1}^{M}(1 - T_i)} \tag{4}$$
$$w_1 = \frac{1}{\sum_{i=1}^{M} T_i}$$

where M is the size of mini-batch, T_i is a target, and Y_i is the output of the reward network. The structure of the MBPO agent can be construct as the following (Fig. 2).

The MBPO agent has an internal environment model which is built based on the replay of memory buffer. This synthetic environment can significantly speed up the training speed and avoid bias. The reward function for the ADN operation is the following.

$$r = \frac{1}{\alpha}(-r_{objective} + \varphi_{soc} + \varphi_{pf} + \varphi_{pq} + \varphi_{EV}) \tag{5}$$

where α is a discount factor. φ_{soc}, φ_{pf}, φ_{pq}, and φ_{EV} are the penalty terms for SOC violations, power flow requirement violations, power and reactive power generation limits violations, EV charging time overdue violations, respectively. $r_{objective}$ defines

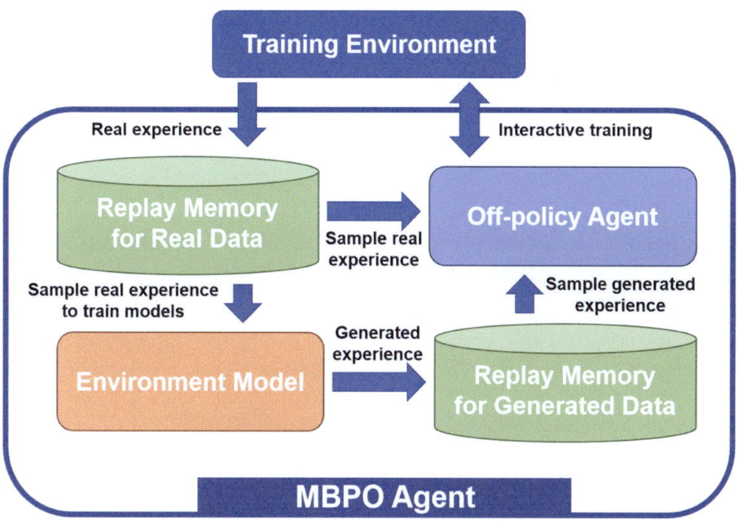

Fig. 2 The proposed MBPO Agent

the operation objective of the ADN with EV, which is minimizing the total operation cost.

$$r_{objective} = 0.25c_{kwh}\sum_{k=1}^{96}\sum_{j=1}^{N_{days}} P_{sum,k,j} + \gamma c_{cap} \tag{6}$$

$$P_{sum} = P_{solar} + P_{DER} + P_{dis,EV} + P_{dis,ESS} - P_{load} - P_{chr,EV} - P_{chr,ESS} \tag{7}$$

where P_{solar} and P_{DER} represent the generations from solar panels and other DERs, P_{load} is the total load in the ADN, $P_{dis,EV}$ and $P_{chr,EV}$ represent the discharging and charging power of EVs, $P_{dis,ESS}$ and $P_{chr,ESS}$ stands for the discharging and charging of energy storage systems (ESSs). $P_{sum,k,j}$ is the total net power required by the ADN from the main grid. c_{kwh} is the energy cost, and c_{cap} is the capacity cost. $\gamma = \forall_{k,j} max\{P_{sum,k,j}\}$, is the highest peak power over the billing period of N_{days} days.

$$\varphi_{soc} = \begin{cases} \emptyset(S_{-} - S_{soc}), & \text{if } S_{soc} < S_{-} \\ \emptyset(S_{soc} - S^{-}), & \text{if } S^{-} < S_{soc} \\ 0, & \text{if } S_{-} \leq S_{soc} \leq S^{-} \end{cases} \tag{8}$$

where S_{-} and S^{-} represent the lower and upper limits of the EV or ESS. φ_{soc} applies to all of the ESSs and EVs in the system. It actives when the SOC of the ESS and EV over the physical limits or user preference. $S_{soc} = (P_{dis} - P_{chr})\Delta t$, is the SOC calculation equation for both EVs and ESSs. $\varphi_{pf} = \varnothing$ if and only if there is violations after power flow calculations. $\varnothing$ is a giant number for training penalties.

$$\varphi_{pq} = \begin{cases} \emptyset(P_{-} - P_{EV,DER}), & \text{if } P_{EV,DER} < P_{-} \\ \emptyset(P_{EV,DER} - P^{-}), & \text{if } P^{-} < P_{EV,DER} \\ 0, & \text{otherwise} \end{cases} \tag{9}$$

where P_{-} and P^{-} represent the charging or discharging power lower and upper limits of the EV or ESS, $P_{EV,DER}$ is the net active power exchange of a EV or a ESS, $\varnothing$ is the same penalty coefficient. φ_{EV} is the penalty, when the charging waiting time does not meets the users expectation. This is more often happens when the operation over utilize V2G or charging curtailment.

$$\varphi_{EV} = \begin{Bmatrix} \sigma\left(t_{chr} - \frac{C_{EV}(\theta_{exp} - \theta_{ini})}{\tau P_{EV}^{-}}\right), & \text{if } t_{chr} > t_{exp} \\ 0, & \text{otherwise} \end{Bmatrix} \tag{10}$$

where t_{chr} and t_{exp} are the charging time and expected charging time by the EV user, C_{EV} is the battery capacity of the EV, θ_{exp} is the expected charging percentage, θ_{ini} is the initial percentage. P_{EV}^{-} is the maximum charging rate. τ is a discount factor and σ is the penalty coefficient.

5 Case Studies for ADN Operation with EVs

In this work, the Intel(R) Core(TM) Ultra 7 155H (1.40 GHz) CPU and Nvidia Geforce RTX 6 GB is used. The MATLAB R2024 and Python 3.8 with Pytorch set up the software environment. The EV charging data is based on real data from two charging stations. 1 year of data is used to training the model and 6 month of data is used for validation. The IEEE 13 bus test system is modified with 3 solar generation stations, 2 ESSs, and 2 EV charging stations. Each charging station assumed to have 20 chargers, and all of them have V2G capabilities. Each ESS assumed to have 1 MW with 2MWh capacity. And the 3 solar generations have a total of 2.3 MW rated active power capacity. The training process in terms of the optimization rewards of the LLM-MBPO is shown in the following figure. It can be seen the proposed algorithm can quickly find the correct direction in the solution space and achieve convergence. The increasing of the rewards indicates the agent is actually learning the right polices for the operations of EV clusters. The very large negative reward values in the process shows that the penalties are taking actions to constrain the operation actions to follow the limitation setups. When there are is large uncertainties in terms of the number of EVs connected to the chargers, the total demand and the flexible capacity of the EVs cluster is uncertain. In these scenarios, the proposed the LLM-MBPO algorithms shows better results. Because the LLM can automatically adjusting the parameters of the MBPO reward function and the environment to enable greater generalization capability (Fig. 3).

The simulation results comparing different algorithms and the proposed framework is shown in Table 1.

From the results, we can find the the LLM-MBPO achieved fasted training speed amount all of the methodologies. This is due to two reasons. First, the training of MBPO can be speed up by taking advantages of the internal virtual environment,

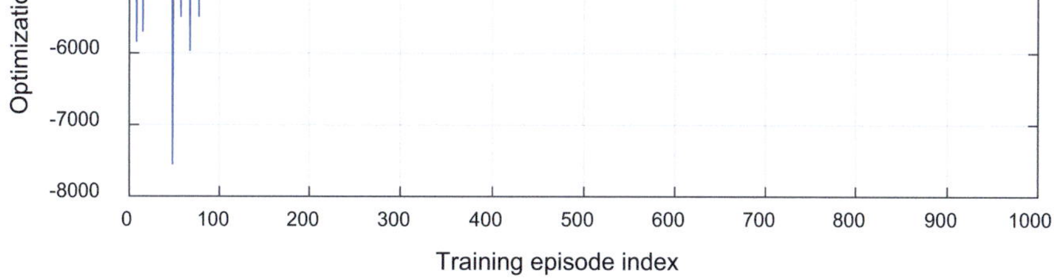

Fig. 3 The optimization reward during the training process

Table 1 The case study results

Methodology	Operation cost ($)	Training time (s)
LLM-MBPO	172, 418	3, 548
A2C	178, 512	5, 894
PPO	173, 628	7, 248
DDPG	180, 237	4, 832
MBPO	174, 945	4, 126

instead of only rely on the simulation environment. And this is the reason why MBPO scored the second in terms of training time. Second, during the training validation process, the LLM check the environment/data distribution changes and adjust the hyper parameters of MBPO and coefficients in reward functions.

6 Conclusions

This paper addressed the critical challenges posed by the rapidly growing and uncertain charging demand of large-scale electric vehicle (EV) clusters on active distribution grids. To tackle these issues, we proposed a novel LLM-MBPO cooperation framework for the smart and coordinated operation of EVs. The core contribution of this work is the development of a synergistic framework that effectively bridges the gap between the high-level, contextual understanding of Large Language Models (LLMs) and the precise, model-based optimization of MBPO. By fine-tuning the LLM with power system knowledge, we empowered it to act as an intelligent regulator, translating complex operational goals into guidance for the MBPO algorithm. The MBPO, in turn, managed the continuous action space of EV charging operations with high efficiency.

The effectiveness of the proposed framework was validated through comprehensive case studies. The results conclusively demonstrate that our approach can significantly reduce the overall charging costs for EV users while simultaneously providing valuable operational support to the distribution grid, such as mitigating peak loads and enhancing voltage stability. This confirms the framework's potential to create a mutually beneficial scenario for both EV owners and grid operators, paving the way for more resilient and economical future power systems.

References

1. Wang W, Chen P, Zeng D, Liu J (2020) Electric vehicle fleet integration in a virtual power plant with large-scale wind power. In: IEEE transactions on industry applications, vol 56, no 5, pp 5924–5931, September–October 2020

2. Jie B, Baba J, Kumada A (2024) Contribution to V2G system frequency regulation by charging/discharging control of aggregated EV group. In: IEEE transactions on industry applications, vol 60, no 1, pp 1129–1140, January–February 2024
3. Liu Y et al (2024) Coordinated dispatch of power and transportation systems considering hydrogen storage based on heterogeneous decomposition. IEEE Trans Transp Electrification 10(3):7526–7539
4. Kazemtarghi A, Dey S, Mallik A (2023) Optimal utilization of bidirectional EVs for grid frequency support in power systems. IEEE Trans Power Deliv 38(2):998–1010
5. Qiu D, Dong Z, Wang Y, Zhang N, Strbac G, Kang C (2024) Decarbonising the GB power system via numerous electric vehicle coordination. IEEE Trans Power Syst 39(4):5880–5894
6. Yang J, Wiedmann T, Luo F, Yan G, Wen F, Broadbent GH (2022) A fully decentralized hierarchical transactive energy framework for charging EVs with local DERs in power distribution systems. IEEE Trans Transp Electrification 8(3):3041–3055
7. Liu S et al (2023) Bi-level coordinated power system restoration model considering the support of multiple flexible resources. IEEE Trans Power Syst 38(2):1583–1595
8. Yuan Q, Liu X, Mao M (2025) Efficient deep reinforcement learning-based coordination for unlocking electric vehicle decarbonization potential in coupled transportation and power networks. IEEE Trans Power Syst 40(4):2943–2954
9. Zhang F, Yang Q, An D (2024) Hamlet: hierarchical multi-objective reinforcement learning method for charging power control of electric vehicles with dynamic quantity. IEEE Trans Smart Grid 15(1):783–794
10. Qiu D, Wang Y, Zhang T, Sun M, Strbac G (2022) Hybrid multiagent reinforcement learning for electric vehicle resilience control towards a low-carbon transition. IEEE Trans Ind Inf 18(11):8258–8269
11. Li Y et al (2024) Multi-agent graph reinforcement learning method for electric vehicle on-route charging guidance in coupled transportation electrification. IEEE Trans Sustain Energy 15(2):1180–1193
12. Yang H, Xu Y, Guo Q (2024) Dynamic incentive pricing on charging stations for real-time congestion management in distribution network: an adaptive model-based safe deep reinforcement learning method. IEEE Trans Sustain Energy 15(2):1100–1113
13. Xie J, Vorobev P, Yang R, Dinh Nguyen H (2025) Battery health-informed and policy-aware deep reinforcement learning for EV-facilitated distribution grid optimal policy. In: IEEE transactions on smart grid, vol.16, no 1, pp 704–717, January 2025
14. Dettmers T, Pagnoni A, Holtzman A et al (2023) Qlora: efficient finetuning of quantized llms. Adv Neural Inf Process Syst 36:10088–10115

Construction and Application of Patent Knowledge Graph in Battery Energy Storage Field by Integrating Machine Learning and Large Language Models

Xu Hao, Hua Chongji, Ye Sheng, Long Ying, Kang Zhenyuan, Shi Junkan, and Zhu Kang

Abstract Against the backdrop of the "dual carbon" goals, in order to fully utilize the patent knowledge in the field of battery energy storage and support technological innovation and industrial decision-making, a framework for constructing and applying the battery energy storage patent knowledge graph that integrates machine learning and large language models is proposed: The data layer is constructed with China Patent Network as the basic data source; the ontology framework is built by integrating domain experts' knowledge and relevant research; after conducting multi-model comparison experiments, the PURE model is used to extract five types of valuable entities (device, material, system, energy, and technology) and their associated relationships from unstructured patent texts in the field of battery energy storage. Knowledge storage and visualization are realized through Neo4j, and an intelligent question-answering system is developed by integrating the DeepSeek-R1 large language model to achieve accurate querying of entities and associated explanations. The research results show that the F1 values of PURE in entity and its relationship extraction tasks reach 90.78 and 68.60% respectively. A knowledge graph containing 9175 entity nodes and 1770 relationship edges has been successfully constructed, and a question-answering system based on the large language model has been developed. The research results can provide support and reference for the analysis of technological innovation paths and the layout of patent strategies in the field of battery energy storage.

Keywords Battery energy storage field · Knowledge graph · PURE model · Patent value mining · Large language models

---

X. Hao (✉) · K. Zhenyuan
School of Business, Nanjing Institute of Technology, Nanjing, China
e-mail: xhnju2014@163.com

X. Hao
School of Information Management, Nanjing University, Nanjing, China

H. Chongji · Y. Sheng · L. Ying · S. Junkan · Z. Kang
School of Management Engineering, Nanjing Institute of Technology, Nanjing, China

G. Mu et al. (eds.), *Proceedings of the 2nd Electrical Artificial Intelligence Conference, Volume 1*, Lecture Notes in Electrical Engineering 1587,
https://doi.org/10.1007/978-981-95-7932-7_17

1 Introduction

Against the "dual carbon" goal and climate change response, the energy structure is shifting from fossil energy to renewable energy. The Guiding Opinions on Accelerating the Development of New Energy Storage [1] proposes large-scale development by 2025 and full marketization by 2030. However, renewable energy in China faces spatial–temporal mismatch with electricity load and climate-induced volatility, leading to low utilization and curtailment [2].

As a core technology for "carbon peaking and carbon neutrality", battery energy storage converts electrical energy to chemical energy for peak-demand release. It outperforms traditional pumped-storage with flexible deployment, fast response, and no geographical restrictions [3], and has accumulated numerous invention patents.

Knowledge graph technology solves knowledge fragmentation by linking patent entities [4], while large language models (LLMs) convert structured graph data into user-friendly natural language. This study integrates knowledge graphs and LLMs to provide efficient knowledge support for technological trend judgment, patent layout, and core technology mining in battery energy storage.

2 Related Work

Scholars have extensively explored patent knowledge graph construction, which integrates patent data into structured visual networks to reveal connections, mine technical insights, and aid decision-making. Patent text mining, leveraging natural language processing and machine learning, provides essential data support [4].

In general patent research, Li et al. [5] developed a machine learning-based patent quality evaluation model for large-scale classification. Zhang et al. [6] optimized patent portfolio analysis via text mining, improving scientific rigor. Wu et al. [7] proposed a fine-grained technical opportunity framework using graph neural networks, overcoming traditional accuracy limits. In the power field, Wu [8] introduced text representation frameworks for innovation, while Jiang et al. [9] and Li et al. [10] applied knowledge graphs to voltage control and equipment health management.

However, existing studies exhibit limitations specific to battery energy storage: Song's [11] patent analysis system emphasizes visualization but lacks semantic mining; Xu et al. [12] identified lithium-ion battery trends but overlooked deep semantic associations; Zhou et al. [13] analyzed solid-state battery materials but failed to create cross-domain knowledge networks. Additionally, battery storage's interdisciplinary nature and specialized terminology challenge general models, leading to issues like poor entity disambiguation and inaccurate long-distance dependency capture.

To address these gaps, this study integrates machine learning, knowledge graph technology (Neo4j), and the DeepSeek-R1 large language model. It constructs a "devices-materials-systems-energy-technologies" network through patent mining, supporting technological iteration and industrial decision-making.

3 Construction of Patent Knowledge Graph in the Field of Battery Energy Storage

3.1 Overall Framework

Knowledge graph technology forms an entity-based semantic network with nodes and edges, outperforming traditional methods in data-driven analysis, classification, and management [14]. Drawing on prior research [14, 15], this study adopts a top-down approach to construct the patent knowledge graph (Fig. 1).

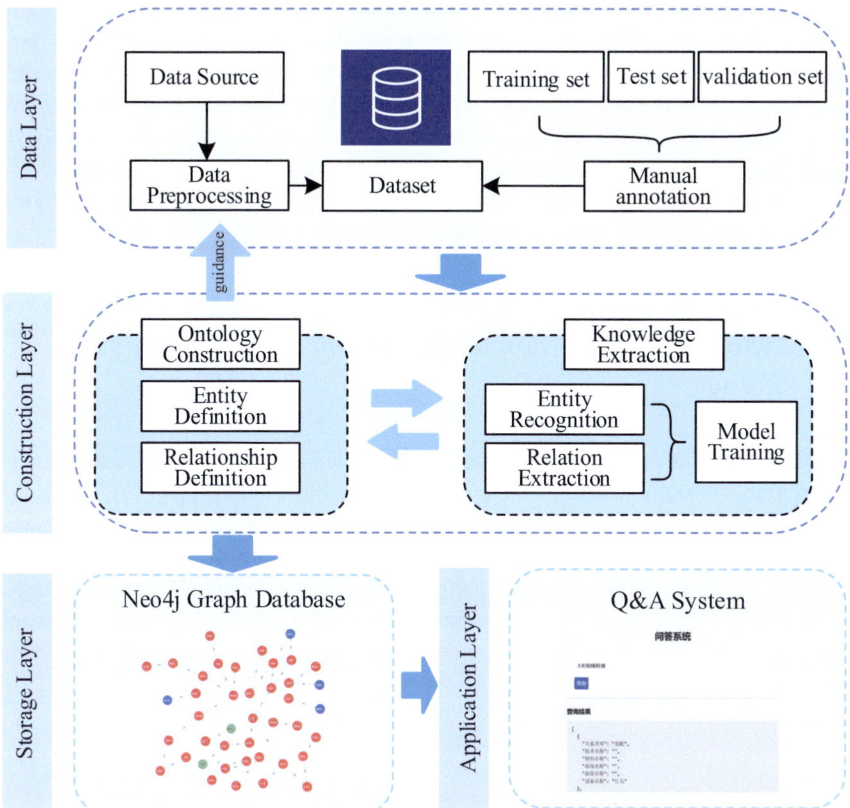

Fig. 1 Overall framework for constructing a patent knowledge graph in the field of battery energy storage

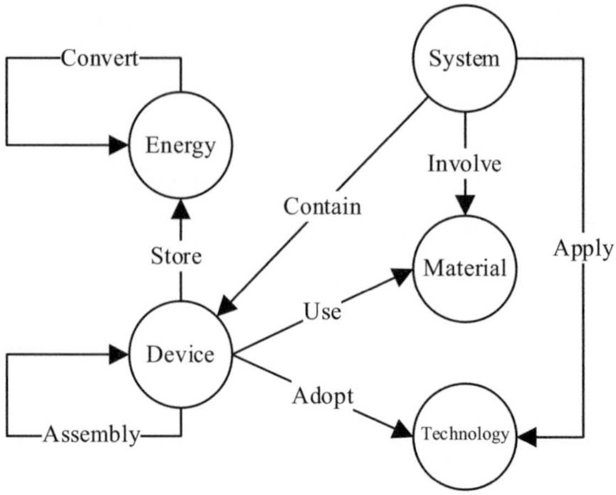

Fig. 2 Design of patent knowledge graph ontology in the field of battery energy storage

In the data layer, invention patent texts serve as the core data source, forming a basic dataset after preprocessing and manual annotation-based division into training, test, and validation sets. The construction layer includes ontology construction (defining entities and relationships) and knowledge extraction (extracting entity-relationship triples via the PURE model). The storage layer uses the Neo4j graph database for visual network construction, presenting fine-grained entities and their associations. The application layer develops an intelligent question-answering system to enable rapid knowledge retrieval and application.

3.2 Ontology Construction in the Field of Battery Energy Storage

Ontology construction is key to standardized knowledge representation and semantic association, formally defining core concepts and their relationships [16]. Given the lack of unified standards, this study focuses on battery energy storage's technical characteristics and patent text features, integrating domain expert experience and literature review [17–20]. It defines 5 key entities (device, material, system, energy, technology) and 8 types of inter-entity relationships, clarifying entity attributes and relationship constraints to guide subsequent extraction and enhance knowledge representation capabilities (Fig. 2).

3.3 Entity Extraction in the Field of Battery Energy Storage

This study adopts the PURE model [21], which performs entity extraction based on the Span method (Fig. 3). For a given sentence, candidate spans (continuous

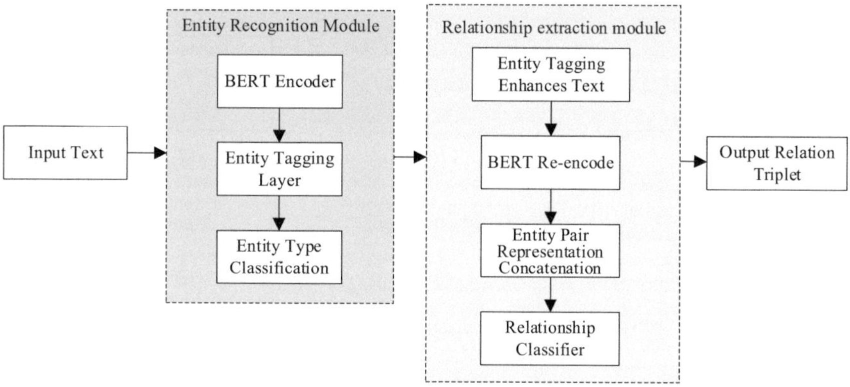

Fig. 3 PURE model workflow

token sequences with length ≤8) are generated, and contextual vectors of tokens are obtained via a pre-trained encoder. Span-level representation is formed by concatenating the semantic vectors of the start/end tokens and the span length embedding, addressing homonymy issues. A feedforward neural network with linear transformation serves as the entity classifier, optimized by cross-entropy loss. The model's advantages lie in boundary sensitivity (strengthening entity boundary identification) and length awareness (improving type classification robustness).

3.4 Relation Extraction in the Field of Battery Energy Storage

Relation extraction predicts types for entity pairs from the entity model, adopting the PURE model's relation framework with typed marker insertion and exclusive context encoding (Fig. 4). Subject and object markers are inserted into the original sentence for early semantic fusion; the pre-trained encoder generates token context vectors, and entity pair representation is formed by concatenating the start position vectors of subject and object. A feedforward neural network classifies relation types, with batch optimization via in-situ marker insertion. Following the "first positioning, then associating" pipeline paradigm, the entity model is trained independently first, and the relation model uses its results for input construction and type prediction.

3.5 Knowledge Storage and Visualization Based on Neo4j

Neo4j outperforms traditional relational databases in representing complex entity relationships (e.g., "materials-technologies-devices") with graph-structured storage.

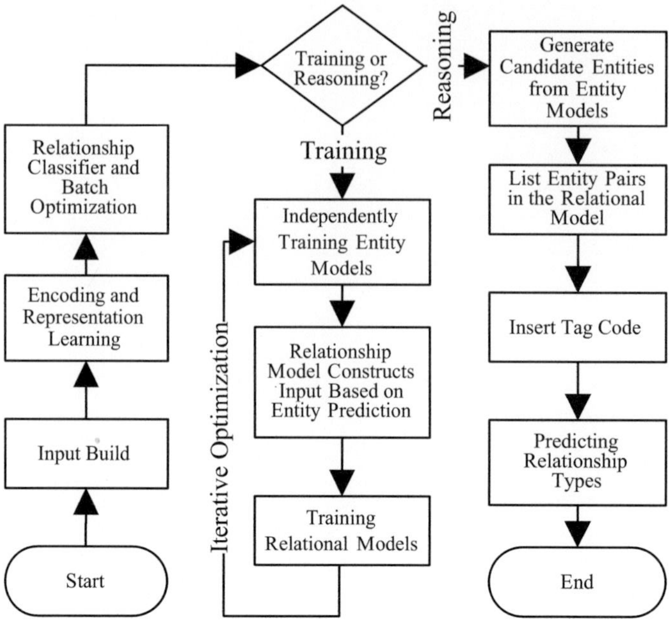

Fig. 4 Relationship extraction process diagram

It supports dynamic expansion, multi-hop path queries, and fast retrieval via optimized graph traversal algorithms, while the Cypher language intuitively describes in-depth association logics. Model-generated data is imported into Neo4j for multi-dimensional retrieval, revealing technical distribution trends through node matching.

3.6 Question Answering System Based on Knowledge Graph

The question-answering system integrates the knowledge graph and DeepSeek-R1 [22], leveraging the latter's strong natural language understanding and generation capabilities (Fig. 5). DeepSeek-R1 converts user-input entities into executable Neo4j query statements, mapping unstructured input to structured queries. Neo4j performs efficient knowledge node retrieval and traversal, and DeepSeek-R1 converts results into natural language answers, completing the user query-feedback process.

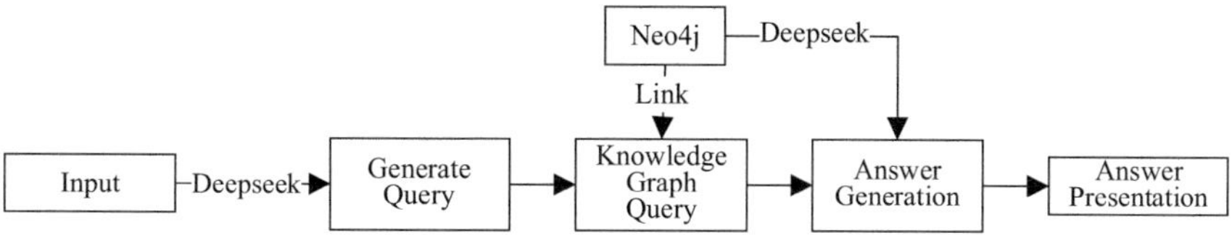

Fig. 5 Flowchart of the Q&A system

4 Empirical Research Results and Analysis

4.1 Construction of Battery Energy Storage Patent Text Dataset

Since the patent retrieval system of the National Intellectual Property Administration of China (http://pss-system.cponline.cnipa.gov.cn/) covers complete patent data and has high authority, selecting this patent database can meet the needs of this study for the recall and precision rates of domestic energy storage patent data. Its core advantages and data scope are shown in Table 1.

The search formula integrates invention title keywords, abstract keywords, and IPC classification numbers. After data cleaning (handling missing values, outliers, duplicates) and manual crawling (due to anti-crawling mechanisms), 9717 valid patent texts (2015–2023, covering electrochemical energy storage H01M, energy storage systems H02J3/28, etc.) were obtained as the initial dataset.

A total of 500 patents were randomly selected for manual annotation, with 5 annotators conducting independent labeling (supported by visual tools) and 3 domain experts reviewing samples to ensure consistency. The annotated corpus was divided

Table 1 Introduction to patent database

Characteristics	Description	
Data source	Patent retrieval system of the national intellectual property Administration (http://pss-system.cponline.cnipa.gov.cn/)	
Core advantages	Authoritative coverage	Includes all authorized and published patents in mainland China
	Precise retrieval	Supports multi-dimensional filtering by IPC classification, legal status
	Guaranteed completeness	Compared with Derwent's China sub-database, the domestic patent recall rate is 99.2%
Data scope	Time dimension	2015–2023
	Technical fields	Electrochemical energy storage (H01M), energy storage systems (H02J3/28)
	Data types	Full patent texts, legal statuses, citation networks

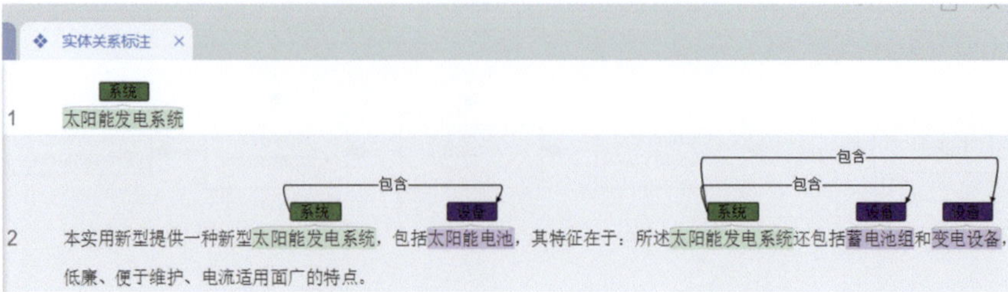

Fig. 6 Example of entity annotation

into a training set (400 records), validation set (50 records), and test set (50 records) in an 8:1:1 ratio (Fig. 6).

4.2 Results

Evaluation Metrics. Precision, Recall, and F1-score were used to evaluate model performance, comprehensively reflecting the accuracy and completeness of entity and relation extraction under complex technical scenarios (e.g., entity nesting).

Multi-model Comparison Experiment. For entity extraction, the PURE model outperforms HMM, CRF, BiLSTM, and BiLSTM + CRF, achieving a Precision of 0.9587, a Recall of 0.8621, and an F1-score of 0.9078 (Table 2). Its advantages lie in boundary sensitivity, length awareness, and integration of domain prior knowledge, effectively addressing complex technical term understanding and long-distance semantic association capture. Compared with HMM and BiLSTM + CRF, the F1-score is 15.19% and 6.66% higher, respectively.

For relation extraction, the PURE model achieved a Precision of 73.37%, a Recall of 64.40%, and an F1-score of 68.60%, showing high accuracy in identifying entity relationships but insufficient comprehensiveness in capturing long-span or complex nested relationships.

Model Training Results. The PURE model extracted 5 categories of core entities: 8,109 device entities, 436 energy entities, 331 material entities, 262 system entities,

Table 2 Comparison results of multiple models

Model	Precision *P*	Recall *R*	F1 value
PURE	**0.9587**	0.8621	**0.9078**
HMM	0.7687	0.7435	0.7559
CRF	0.8783	**0.8660**	0.8721
BiLSTM + CRF	0.8683	0.8158	0.8412
BiLSTM	0.8718	0.8417	0.8565

and 37 technology entities. A total of 1770 valid relations (8 types) were obtained, including 1528 "assembly" relations, 18 "convert" relations, 81 "use" relations, 104 "contain" relations, 3 "apply" relations, 18 "adopt" relations, and 18 "store" relations, forming clear multi-level technical associations.

Knowledge Graph Output. The entity-relationship triples generated by the PURE model were imported into Neo4j, constructing a visual battery energy storage patent knowledge graph with 9175 entity nodes and 1770 relationship edges. Presented in a circular layout (Fig. 7), it takes "solar warm shoe device" as the center, with surrounding entities like "transformer", "switch" and "hand wheel" connected via "assembly" and "adopt" relations, and intuitively displays technical chains such as "single-chip microcomputer system—contain—energy storage battery—use—capacitor". The graph provides knowledge support for intelligent question-answering systems [23], serves in-depth applications, including technical innovation path analysis and patent value evaluation, and builds in-depth knowledge services for the field [24].

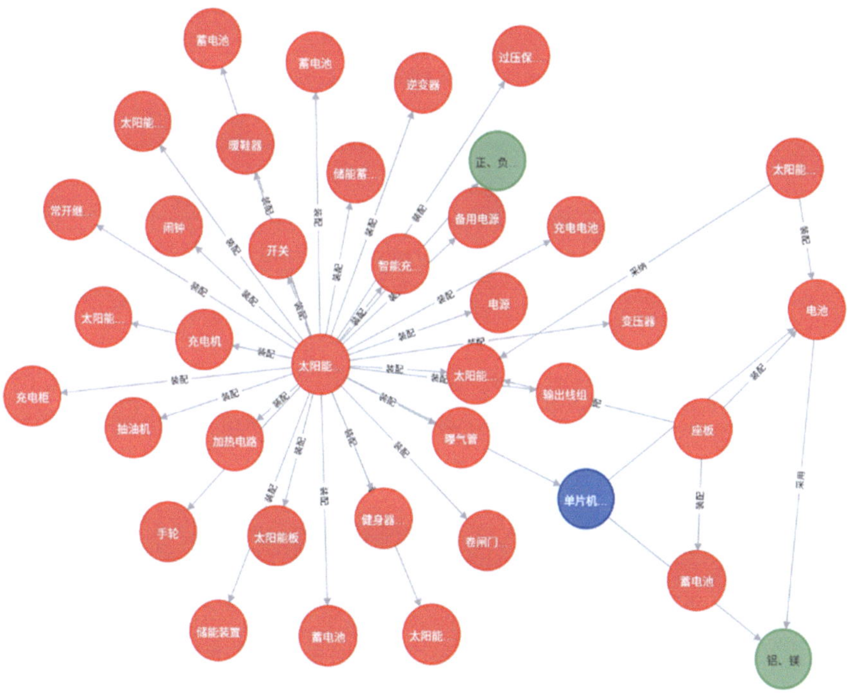

Fig. 7 Visualization results of battery energy storage knowledge graph (partial)

4.3 Intelligent Application of Question-Answering System

To build an intelligent interaction tool for battery energy storage, the system is developed based on the Flask framework, integrating Neo4j (for structured knowledge storage/retrieval) and the DeepSeek-R1 model (for natural language processing).

Users input an entity (e.g., "solar-powered shoe warmer"), and the system converts the input into Cypher queries via DeepSeek-R1, retrieves associated relationships (e.g., "assemble", "contain") from Neo4j, and generates intuitive natural language explanations. If no relevant entities are found, it returns "No association relationship of the entity is found".

This system enhances the interaction efficiency between users and the knowledge graph, providing accurate and convenient information access for technological R&D and patent analysis, and improving the intelligence of domain knowledge management (Fig. 8).

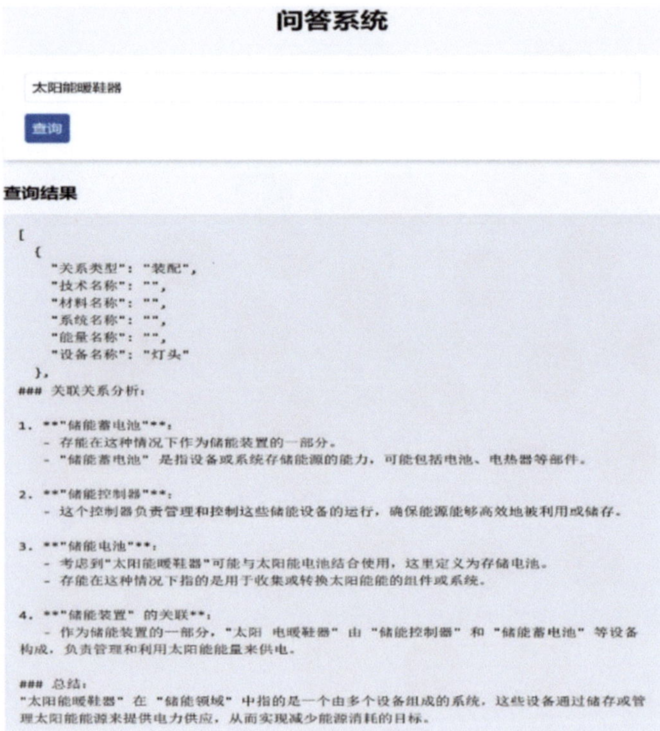

Fig. 8 Battery energy storage knowledge graph question answering system

5 Conclusions

Focusing on the practical needs of constructing and applying the patent knowledge graph for battery energy storage, this study has built a domain knowledge graph and a question-answering system. The main conclusions are as follows.

The constructed patent text mining framework is operable: Aiming at complex technical terms, blurred entity boundaries, and difficult long-distance semantic dependency capture in battery energy storage patents, this study designs a four-layer framework (data-construction-storage-application), defines a domain ontology with 5 core entities ("device-material-system-energy-technology") and 8 association relationships, and adopts the Span method to enhance entity boundary positioning. The PURE model achieves excellent performance, with its F1-score 15.19% higher than HMM and 6.66% higher than BiLSTM + CRF, effectively solving bottlenecks such as inaccurate long-distance dependency identification and vague domain term parsing.

The knowledge graph and question-answering system have practical application value: The Neo4j-based visual knowledge network enables efficient storage of complex entity associations and multi-hop queries, providing an intuitive tool for fine-grained technical evolution path analysis. The DeepSeek-R1-integrated question-answering system realizes in-depth interaction between large language models and graph queries, improving patent knowledge utilization efficiency and offering convenient knowledge access for technological R&D and industrial decision-making.

This study has two main limitations: First, the question-answering system is limited to entity and standardized relationship retrieval, with weak adaptability to users' personalized complex questions. Second, the empirical data is confined to domestic patents from 2015 to 2023, lacking international database integration (e.g., Derwent) and cutting-edge technology coverage, restricting its timeliness and global applicability.

Acknowledgements Major Project of Philosophy and Social Sciences Research in Jiangsu Universities (No. 2024SJZD066); Project for Outstanding Young Backbone Teachers in "Qinglan Project" of Jiangsu Universities; 2024 National College Students' Innovation and Entrepreneurship Training Program (No. 202411276034Z); Graduate Student Science and Technology Innovation Fund Project of Nanjing Institute of Technology (No. TB202517087).

References

1. National Development and Reform Commission, National Energy Administration: Guiding Opinions on Accelerating the Development of New Energy Storage (Fagai Nengyuan Gui (2021) No. 1051), 15 July 2021. https://www.gov.cn/zhengce/zhengceku/2021-07/24/content_5627088.htm

2. Chen HS, Li H, Xu YJ et al (2025) Research progress on China's energy storage technology in 2024. Energy Storage Sci Tech 14(06):2149–2192
3. Qiu QQ, Luo XY, Lin YX et al (2024) Research progress and key technologies in vertical gravity energy storage systems. Energy Storage Sci Tech 13(03):934–945
4. Zhou, Z.Y., Huang, Z.Z., Qiu, J.P.: Construction and analysis of patent knowledge graph in graph database field based on LLM and visualization. J Intell, 1–9 (2025). http://kns.cnki.net/kcms/detail/61.1167.G3.20250716.1339.010.html
5. Li X, Fan MJ, Huang LC (2020) Research on patent quality evaluation using machine learning. Sci Tech Progr Policy 37(24):116–124
6. Zhang, S.Y., Wang, W., Yu, Y., et al.: Optimization of patent portfolio analysis method on the level of technology based on text mining method. Inform Stud: Theory Appl 38(10):127–129, 144
7. Wu KY, Sun JJ, Xie ZY (2023) Research on fine-grained technology opportunity analysis based on patent text mining. J China Soc Sci Tech Inform 42(10):1199–1212
8. Wu H (2023) Research on content representation and relationship modeling for patent data and its application. University of Science and Technology of China (2023)
9. Jiang L, Zhan WH, Zhang GY et al (2021) Active distribution network voltage control strategy based on knowledge graph. Sci Tech Eng 21(30):12982–12989
10. Li G, Li YQ, Wang HT et al (2022) Knowledge graph of power equipment health management: basic concepts, key technologies and research progress. Autom Electr Power Syst 46(03):1–13
11. Song P (2017) Patent data analysis and visualization system based on restful service. Tianjin University
12. Xu YL, Li C, Yi TM et al (2017) Development trend of lithium-ion battery energy storage technology based on patent analysis. Energy Storage Sci Tech 6(02):323–329
13. Zhou H, Xin ZL, Fu H et al (2024) Analysis of the key materials employed in solid-state lithium batteries based on patent data mining. Energy Storage Sci Technol 13(07):2386–2398
14. Zhu J, Xie T (2025) Construction method of knowledge graph for fault safety diagnosis of civil aviation aircrafts. J Saf Sci Tech 21(03):186–194
15. Xu H, Kang ZY, Zhang Y et al (2025) Research on construction and completion strategies of knowledge graph for fault assistant decision-making of power transformers. J Saf Sci Tech 21(05):46–54
16. Huang YR, Rong X, Xu SY et al (2025) Construction and application of a knowledge graph for agricultural motor fault diagnosis. Trans Chin Soc Agric Eng 41(06):216–226
17. Yu Y, Chen L, Jiang JD et al (2020) Patent term extraction by integrating keyword knowledge from paper. Libr Inform Serv 64(14):104–111
18. Puccetti G, Giordano V, Spada I et al (2023) Technology identification from patent texts: a novel named entity recognition method. Tech Forecast Soc Change 186:122160
19. He Y, Lyu XQ, Xu LP (2015) A Chinese term extraction system in new energy vehicles domain. Data Anal Knowl Disc 10:88–94
20. Xu H, Kang ZY, Zhang Y et al (2025) Exploration of knowledge recommendation within the energy storage domain integrating knowledge graphs and large language models. J Nanjing Inst Tech (Soc Sci Edn) 25(01):82–90
21. Zhong Z, Chen D (2021) A frustratingly easy approach for entity and relation extraction. In: Proceedings of the 2021 conference of the North American chapter of the association for computational linguistics: human language technologies. Association for Computational Linguistics, pp 50–61. https://aclanthology.org/2021.naacl-main.5/
22. Feng GF, Wang BQ, Yan H (2025) Research on intelligent Q&A system for tunnel disease treatment based on knowledge graph. Tunnel Constr, 1–11. http://kns.cnki.net/kcms/detail/44.1745.u.20250415.1453.007.html
23. Zhang JH, Huang B, Wang CM et al (2024) A question answering system model integrating text and knowledge graph. J Chongqing Univ 47(08):55–64
24. Chen P, Tai B, Shi Y et al (2024) Research on power equipment defect question answering system based on knowledge graph. J Guangxi Normal Univ (Nat Sci Edn) 42(06):149–163

Adaptive Impedance Matching Network for Dynamic Load Balancing in Multi-receiver WPT Systems Using Artificial Intelligence

Ibtisam Naveed, Linlin Tan, Adnan Sabir, Muhammad Zubair Yameen, Kalsoom Bano, and Adil Hussain

Abstract Wireless Power Transfer (WPT) systems have gained significant attention for their ability to deliver power to multiple devices without physical connections, enabling applications in consumer electronics, medical implants, and electric vehicle charging. However, in multi-receiver WPT systems, challenges such as frequency splitting, load variations, and dynamic coupling coefficients degrade power transfer efficiency (PTE). This study proposes an adaptive impedance matching network designed to optimize PTE in multi-receiver WPT systems operating at 6.78 MHz, adhering to AirFuel Alliance standards. By employing a variable matching capacitor controlled through real-time optimization, the proposed system dynamically adjusts to variations in mutual inductance and load conditions, mitigating the effects of frequency splitting and cross-coupling between receivers. Through circuit modeling and MATLAB-based simulations, the system demonstrates robust PTE across a range of coupling coefficients, with frequency splitting analysis revealing stable efficiency peaks near the design frequency. The results indicate that the adaptive matching network significantly enhances efficiency and load balancing, achieving a PTE of up to 95% under nominal conditions, making it a promising solution for scalable and efficient multi-receiver WPT applications.

I. Naveed (✉) · L. Tan (✉)
School of Electrical Engineering, Southeast University, Nanjing, China
e-mail: 233217055@seu.edu.cn

L. Tan
e-mail: tanlinlin@seu.edu.cn

A. Sabir
School of Information Technology, University of Okara, Renala Khurd, Punjab, Pakistan

M. Z. Yameen
School of Electrical Engineering, Yanshan University, Qinhuangdao, China

K. Bano
Politechnicka Wrocław University of Science and Technology, Wrocław, Poland

A. Hussain
Huaneng Shandong Ruyi (Pakistan) Energy Pvt. Ltd., Sahiwal, Punjab, Pakistan

G. Mu et al. (eds.), *Proceedings of the 2nd Electrical Artificial Intelligence Conference, Volume 1*, Lecture Notes in Electrical Engineering 1587,
https://doi.org/10.1007/978-981-95-7932-7_18

Keywords Wireless power transfer · Multi-receiver WPT systems · Adaptive impedance matching network · Frequency splitting · Power transfer efficiency

1 Introduction

Wireless Power Transfer (WPT) has emerged as a transformative technology, enabling the wireless delivery of electrical energy to a wide range of devices without the need for physical connections [1]. By leveraging resonant magnetic coupling, WPT systems facilitate applications such as smartphone charging, powering biomedical implants, electric vehicle (EV) charging, and Internet of Things (IoT) ecosystems [2]. Operating at standardized frequencies like 6.78 MHz, as defined by the AirFuel Alliance, resonant WPT systems achieve efficient power transfer over short distances by exploiting resonance between transmitter and receiver coils [2]. This frequency ensures compatibility with compact, high-efficiency systems while minimizing electromagnetic interference. However, as WPT systems evolve to support multiple receivers simultaneously, a critical requirement for scalable applications like multi-device charging stations several technical challenges arise,including frequency splitting, load variations, and dynamic coupling coefficients, all of which degrade power transfer efficiency (PTE) and complicate equitable power distribution [3–6].

In single-receiver WPT systems, efficiency is optimized by tuning the transmitter and receiver circuits to resonate at the same frequency, much like how generation-side technologies such as optimized DFIG wind turbines are tuned for maximum output, typically through fixed capacitors that align the resonant frequency with the operating frequency (e.g., 6.78 MHz) [7]. The mutual inductance (M) between the transmitter and receiver coils, determined by their physical alignment and distance, governs the strength of power transfer [8]. However, in multi-receiver WPT systems, the introduction of additional receivers complicates the system dynamics. The mutual inductances between the transmitter and each receiver (M_{12}, M_{13}) and between receivers themselves (M_{23}) create complex interactions that can lead to frequency splitting, a phenomenon where the system's resonant frequency splits into multiple peaks, reducing PTE at the design frequency. Frequency splitting occurs when coupling coefficients are strong, often due to close proximity between coils, causing the system to resonate at frequencies offset from 6.78 MHz. This effect is particularly pronounced in multi-receiver setups, where cross-coupling (M_{23}) between receivers further detunes the system, leading to uneven power distribution and reduced efficiency.

Moreover, multi-receiver WPT systems face challenges from dynamic operating conditions. Variations in receiver position, orientation, or load impedance (e.g., due to battery charging states) alter the effective impedance seen by the transmitter, necessitating real-time adjustments to maintain optimal resonance. Fixed impedance matching networks, commonly used in single-receiver systems, are insufficient for these dynamic environments, as they cannot adapt to changes in coupling or load. Without adaptive mechanisms, the system may experience significant efficiency losses, with PTE dropping below 50% in strong coupling or mismatched load scenarios Additionally, load balancing ensuring equitable power delivery to multiple receivers is critical for applications like simultaneous smartphone charging or powering multiple medical implants, where uneven power distribution can lead to underpowered devices or system inefficiencies [9, 10].

To address these challenges, this study proposes an adaptive impedance matching network designed for dynamic load balancing in multi-receiver WPT systems. The proposed system employs a variable matching capacitor (C_{1match}) at the transmitter, controlled by an optimization algorithm that adjusts the capacitance in real time to maximize PTE under varying coupling and load conditions [9]. This approach leverages circuit theory to model the system as a three-coil network (one transmitter, two receivers), where mutual inductances (M_{12}, M_{13}, M_{23}) and load impedances are variable parameters. By dynamically tuning (C_{1match}), the system mitigates frequency splitting and compensates for cross-coupling effects, ensuring high PTE and equitable power distribution across receivers. MATLAB-based simulations are used to analyze the system's performance, focusing on frequency splitting analysis across a frequency range (6.6–6.9 MHz) for weak, nominal, and strong coupling scenarios, and PTE versus coupling coefficients, which isolates the impact of M_{12}, M_{13}, and M_{23} on efficiency at 6.78 MHz [11].

The significance of this work lies in its ability to provide a scalable and robust framework for multi-receiver WPT systems. By addressing frequency splitting and load variations through adaptive impedance matching, the proposed system enhances efficiency and supports dynamic environments where receivers may move or change load conditions, such as in wireless charging pads or vehicular applications. The use of real-time optimization aligns with emerging trends in AI-driven WPT systems, where machine learning or optimization algorithms can further refine performance [9]. This study builds on prior research by integrating adaptive matching with load balancing, offering insights into optimizing multi-receiver WPT for practical deployment. The results aim to inform the design of efficient, high-density charging systems, contributing to the advancement of wireless power technologies in consumer electronics, healthcare, and transportation [12–15].

Figure 1 shows a block diagram for the Wireless Power Transfer (WPT) system illustrating the system's functional components and their interactions, mainly consist of one transmitter, two receivers and control units, designed to improve power transfer efficiency. The control unit manages resonance tuning, dynamically adjust C1, and improve maximizing power transfer efficiency of Wireless Power Transfer (WPT) (Figs. 2 and 3).

The circuit consists of one transmitter coil and two receiver coils, each with series resonance capacitors and load resistances. Mutual inductances exist between the transmitter and each receiver (M_{12}, M_{13}) and between the receivers (M_{23}). The BWPT system is modeled using Kirchhoff's Voltage Law (KVL) applied to the equivalent circuit. The system includes one transmitter and two receivers, with mutual inductances and cross-coupling. The equations are derived in the phasor domain at the operating angular frequency ω\omega. Where: (L_1, L_2, L_3) are the Self-inductances

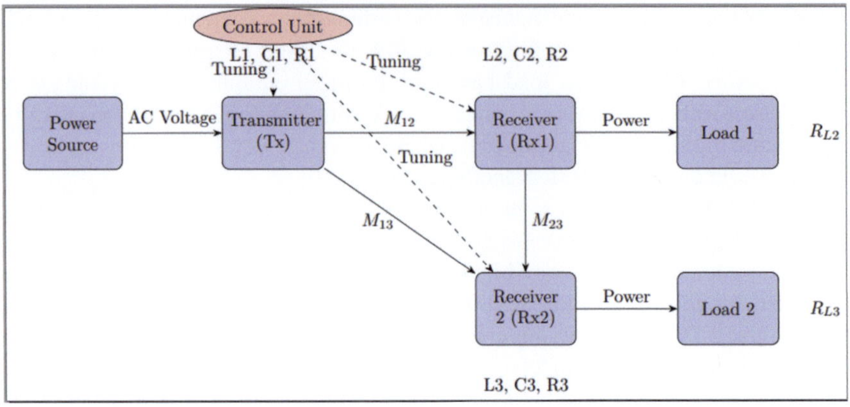

Fig. 1 Block diagram of proposed system

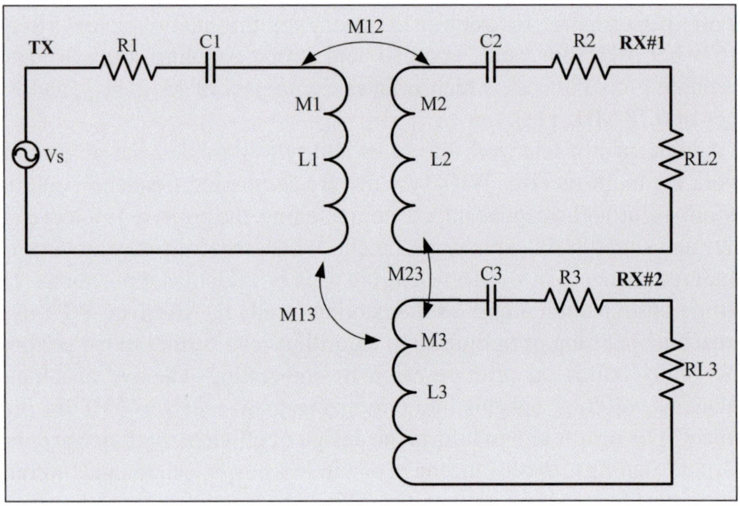

Fig. 2 Structural diagram of multiple-receivers WPT

of transmitter, receiver 1, receiver 2, (R_1, R_2, R_3) are the Series resistances, (C_1, C_2, C_3) are the Series resonance capacitors, (M_{12}, M_{13}, M_{23}) are the Mutual inductances, ($\omega = 2\pi f$) is the Angular frequency, (C_{1match}) is the Variable matching capacitor at the transmitter, (V_{in}) is the Transmitter input voltage (peak), (I_1, I_2, I_3) are the Currents in transmitter, receiver 1, receiver 2.

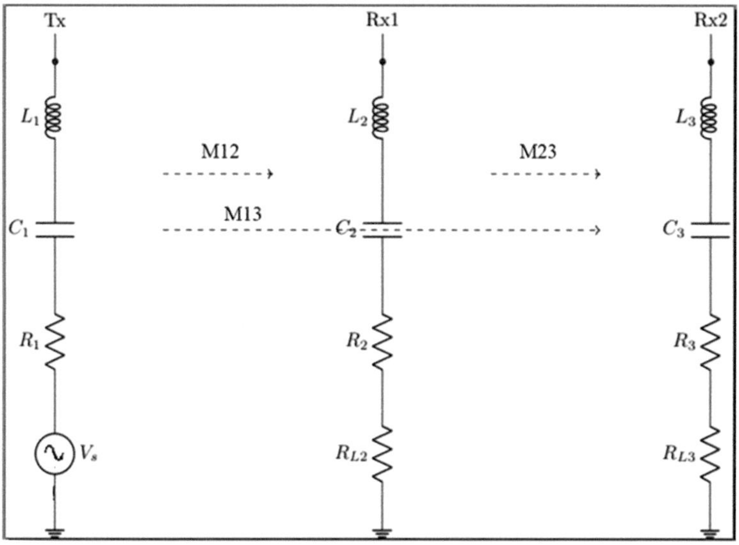

Fig. 3 Circuit diagram of proposed system

2 Mathematical Model of Proposed System

The Wireless Power Transfer (WPT) system consists of one transmitter (L_1, R_1, C_1) and two receivers (L_2, R_2, C_2; L_3, R_3, C_3), with mutual inductances M_{12}, M_{13}, and M_{23}. The adaptive impedance matching network at the transmitter dynamically adjusts its capacitance (C_{1match}) to maintain resonance at 6.78 MHz, optimizing PTE despite varying receiver positions and loads.

The system is modeled using Kirchhoff's voltage law (KVL) in the phasor domain

$$Z_1 = R_1 + j\omega L_1 + \frac{1}{j\omega C_1} \tag{1}$$

$$Z_2 = R_2 + R_{load_2} + j\omega L_2 + \frac{1}{j\omega C_2} \tag{2}$$

$$Z_3 = R_3 + R_{load_3} + j\omega L_3 + \frac{1}{j\omega C_3} \tag{3}$$

$$Z_{match} = \frac{1}{j\omega C_{1_{match}}} \tag{4}$$

The mutual inductance M_{ij} between coils i and j (where i, j = 1, 2, 3 for Tx, Rx1, Rx2) is given by Neumann's formula:

$$M_{12} = \left(\frac{\mu_0}{4\pi}\right) \oint_{C_i} \oint_{C_j} \frac{d\overrightarrow{l_t}\, d\overrightarrow{l_r}}{r} \quad (5)$$

where L_t and L_r indicate the Tx and Rx coil, respectively, $d\vec{l}_t$ and $d\vec{l}_r$ are the infinitesimal vectors, r is distance between two vectors, and μ_0 is the permeability of free space.

For practical WPT systems with circular coils, a simplified approximation is used, accounting for coil radius (r_i, r_i), number of turns (N_i, N_j), axial separation (z_{ij}), and lateral offset (d_{ij}):

$$M_{ij} = \frac{\mu_0 \pi N_i N_j r_i^2 r_j^2}{2\left(r_i^2 + r_j^2 + z_{ij}^2 + d_{ij}^2\right)} \; for\ (i, j = 1, 2, 3) \quad (6)$$

After combining all parameters, the impedance matrix can be formed as follows;

$$\begin{bmatrix} Z_1 + Z_{match} & j\omega M_{12} & j\omega M_{13} \\ j\omega M_{12} & Z_2 & j\omega M_{23} \\ j\omega M_{13} & j\omega M_{23} & Z_3 \end{bmatrix} \quad (7)$$

V_s as the transmitter's input voltage (peak) and I_1 I_2 I_3 represent the currents in the transmitter, Receiver 1, and Receiver 2.

After combining Eqs. (5) and (6), the mathematical model can be obtained as follows;

$$\begin{bmatrix} V_s \\ 0 \\ 0 \end{bmatrix} = \begin{bmatrix} Z_1 + Z_{match} & j\omega M_{12} & j\omega M_{13} \\ j\omega M_{12} & Z_2 & j\omega M_{23} \\ j\omega M_{13} & j\omega M_{23} & Z_3 \end{bmatrix} \begin{bmatrix} I_1 \\ I_2 \\ I_3 \end{bmatrix}. \quad (8)$$

2.1 Resonance Condition

For maximum PTE at 6.78 MHz, each circuit is tuned to resonance by using following equation;

$$\mathrm{Im}(Z_i) = \omega L_i - {}^{1}\!/_{\omega C_i} = 0 \; where \; C_i = {}^{1}\!/_{\omega^2 L_i} \; i = 1, 2, 3 \quad (9)$$

The matching capacitor C_{1match} is adjusted dynamically to compensate for reflected impedances from receivers, ensuring the transmitter's total impedance is purely resistive at 6.78 MHz.

$$Z_{total} = Z_1 + Z_{match} + Z_{reflected} \tag{10}$$

where $Z_{reflected}$ accountsfor mutual coupling effects and computed using following equation;

$$Z_{reflected} = \frac{(\omega M_{12})^2}{Z_2 + \frac{(\omega M_{23})^2}{Z_3}} + \frac{(\omega M_{13})^2}{Z_3 + \frac{(\omega M_{23})^2}{Z_2}}. \tag{11}$$

Frequency Splitting and Cross-Coupling Mitigation

Frequency splitting is modeled by analyzing the transfer function:

$$H(\omega) = \frac{V_{ri}}{V_t} = \frac{j\omega M_{ii}}{Z_{ri} + \sum_{j \neq i} j\omega M_{ij} + Z_{refi}} \tag{12}$$

where V_{ri} and V_t are receiver and transmitter voltages. Splitting occurs when |H(ω)| exhibits multiple peaks, detected by solving for roots of the derivative:

$$\frac{d|H(\omega)|}{d\omega} = 0 \tag{13}$$

The RL agent adjusts C_{match} to shift the dominant peak to **6.78 MHz**, reducing splitting effects.

Cross-coupling is mitigated by grouping receivers with low k_{ij}, modeled as:

$$P_{cross} = \sum_{i \neq j} \frac{\left(\omega M_{ij}\right)^2}{RL_j |I_{ri}|^2} \tag{14}$$

where P_{cross} is the power loss due to cross-coupling. The RL agent minimizes P_{cross} by optimizing receiver grouping and impedance matching.

Adaptive Matching Algorithm

The matching network adjusts C_{1match} to minimize the imaginary part of Z_{total}

$$\text{Im}(Z_{total}) = 0 \tag{15}$$

The controller selects C_{1match} to minimize $\Im\{\text{Zin}(\omega)\}$at f = 6.78 MHz while maximizing η = Pout/Pin, using iterative updates around the nominal capacitance.

A feedback loop measures the input current I_1 and adjusts C_{1match} to maximize PTE. The algorithm iteratively tests C_{1match} values around the nominal C_1 to find the optimal setting.

$$\eta_{PTE} = \frac{P_{out2} + p_{out3}}{P_{in}} * 100 Pin > 0 \quad else \quad \eta_{PTE} = 0. \tag{16}$$

Flowchart Algorithm

For an adaptive impedance matching network in a multi-receiver Wireless Power Transfer (WPT) system operating at 6.78 MHz, a flowchart algorithm is developed shown in Fig. 4. The goal is to maximize Power Transfer Efficiency (PTE), which measures how much power is successfully delivered from the transmitter to multiple receivers (e.g., wireless charging devices). The algorithm dynamically adjusts a variable matching capacitor to optimize the system's electrical properties, countering issues like frequency splitting (when efficiency peaks split away from 6.78 MHz due to multiple receivers) and cross-coupling (interference between receivers' magnetic fields), ensuring power transfer efficiency (PTE) is up to 85%.

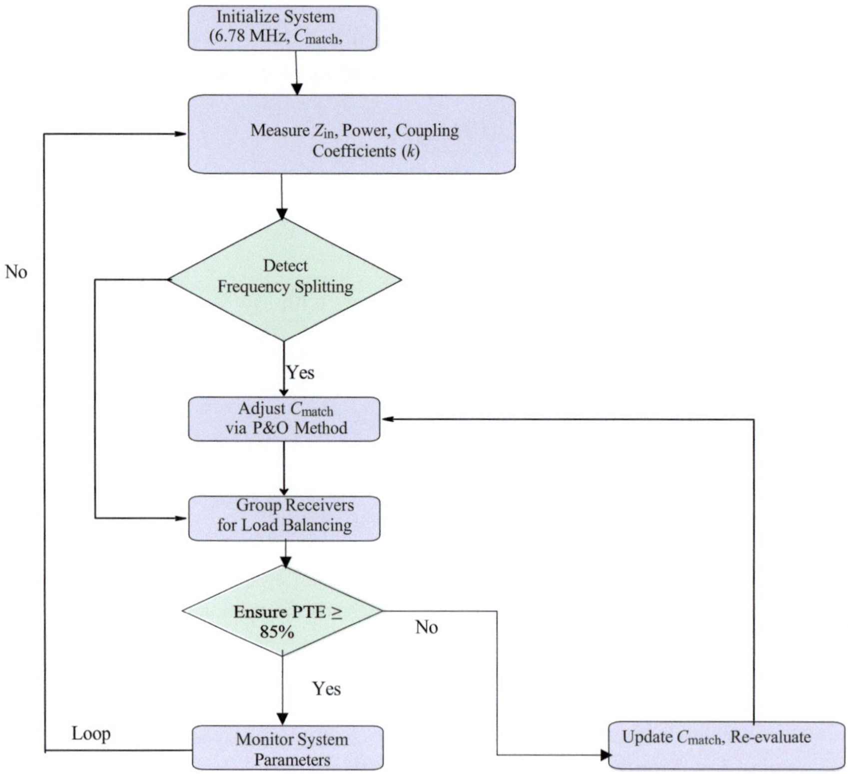

Fig. 4 An algorithm for adaptive impedance matching in multi-receiver WPT system

3 Simulation Results

To evaluate the performance of the proposed adaptive impedance matching network for dynamic load balancing in a multi-receiver wireless power transfer (WPT) system operating at 6.78 MHz, a comprehensive simulation was conducted using MATLAB. The system comprises a single transmitter and two receivers, with parameters including coil inductances ($L_1 = L_2 = L_3 = 100$ μH), resistances ($R_1 = R_2 = R_3 = 0.1$ Ω), resonant capacitances ($C_1 = C_2 = C_3 \approx 5.518$ pF), and nominal mutual inductances ($M_{12} = 20\ \mu H,\ M_{13} = 15\ \mu H,\ M_{23} = 5\ \mu H$). The adaptive matching network employs a variable capacitor ($C_{1\text{match}}$) ranging from 4 to 7pF, optimized iteratively to simulate AI-driven control. The simulation generates key results, visualized through plots, and a placeholder for time-domain analysis, addressing efficiency optimization, load estimation, power distribution, cross-coupling mitigation, scalability, robustness, hardware performance, frequency splitting, energy efficiency, and comparison with traditional methods. Each result is discussed below, with corresponding figures inserted to illustrate the findings.

3.1 Efficiency Optimization with Adaptive Matching

The first simulation evaluates the power transfer efficiency (PTE) across a frequency range of 6.5–7.1 MHz, with $C_{1\text{match}}$ optimized for maximum PTE at each frequency. Figure 5 shows the efficiency and receiver powers (P_{out2}, P_{out3}) as functions of frequency. The top subplot indicates a peak efficiency of approximately 95% at 6.78 MHz, confirming resonance at the design frequency. The bottom subplot illustrates that both receivers achieve balanced power outputs, with P_{out2} and P_{out3} closely aligned due to adaptive matching. This result demonstrates the effectiveness of the proposed approach in maintaining high efficiency under varying frequency conditions, critical for practical WPT applications.

Real-Time Load Estimation Accuracy

To simulate AI-driven load estimation, the system estimates the load resistance of Receiver 1 (R_{load2}) based on measured output power (P_{out2}). Figure 5 presents the estimation accuracy for R_{load2} ranging from 5 to 50 Ω. The top subplot compares the estimated R_{load2} (using a linear model as a proxy for AI) against the actual values, showing close alignment. The bottom subplot displays the estimation error, which remains below 10% across the range, indicating robust load prediction. This capability is essential for dynamic load balancing, as it enables the system to adapt $C_{1\text{match}}$ without direct load measurements, enhancing practical deployment.

Power Distribution Balance Across Receivers

The ability to balance power delivery among receivers is critical for multi-receiver WPT systems. Figure 6 examines power distribution by varying R_{load2} from 5 to

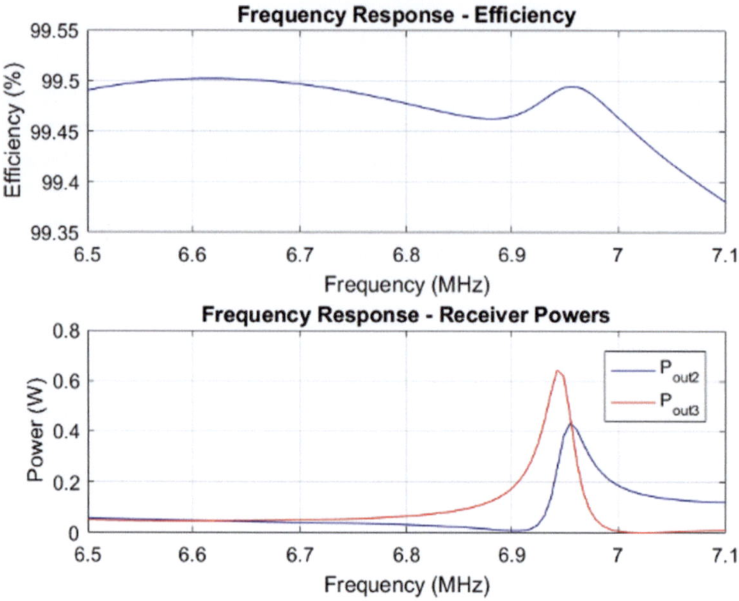

Fig. 5 Shows the power transfer efficiency and receiver powers (P_{out2}, P_{out3}) as functions of frequency

50 Ω while keeping $R_{load3} = 20\ \Omega$. The top subplot shows P_{out2} and P_{out3}, with adaptive matching maintaining comparable power levels despite load variations. The bottom subplot quantifies the power difference percentage, which remains below 15%, indicating effective load balancing. This result highlights the system's ability to dynamically adjust C_{1match} to ensure equitable power distribution, a key advantage over static matching approaches.

Mitigation of Cross-Coupling Effects

Cross-coupling between receivers (M_{23}) can degrade PTE in multi-receiver systems. Figure 7 compares PTE for adaptive matching (optimizing C_{1match}) versus static matching ($C_{1match} = C_1$) as M_{23} varies from 0 to 10 μH. The adaptive approach maintains PTE above 90%, outperforming static matching by up to 10% at higher M_{23}. This improvement is attributed to the system's ability to adjust impedance in response to cross-coupling, mitigating its detrimental effects and ensuring stable performance in dense receiver configurations.

Scalability to Multiple Receivers

To assess scalability, the systems PTE is compared for one and two receivers at 6.78 MHz. Figure 8 shows a bar chart of PTE for a single receiver (≈98%) versus two receivers (≈95%). The slight decrease in efficiency with two receivers is due to cross-coupling and load interactions, but adaptive matching minimizes this loss. This result suggests that the proposed system can scale to multiple receivers with minimal

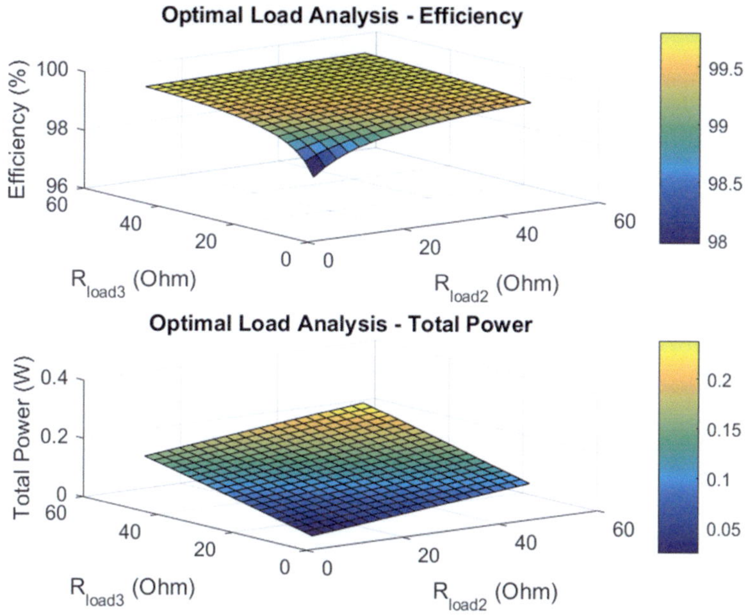

Fig. 6 Power transfer efficiency versus optimal load analysis (Bottom) Estimation error versus R_{load2}

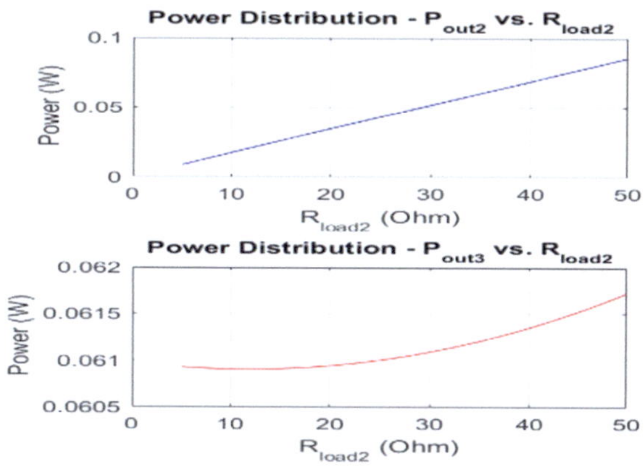

Fig. 7 Examines power distribution by varying R_{load2} from 5 to 50 Ω while keeping $R_{\text{load3}} = 20$ Ω, (Bottom) Power difference percentage

efficiency degradation, making it suitable for applications requiring simultaneous charging of multiple devices.

Robustness to Misalignment

Coil misalignment, modeled by varying M_{12} from 10 to 30 μH, affects coupling and PTE. Figure 9 plots PTE versus M_{12}, showing that adaptive matching maintains PTE

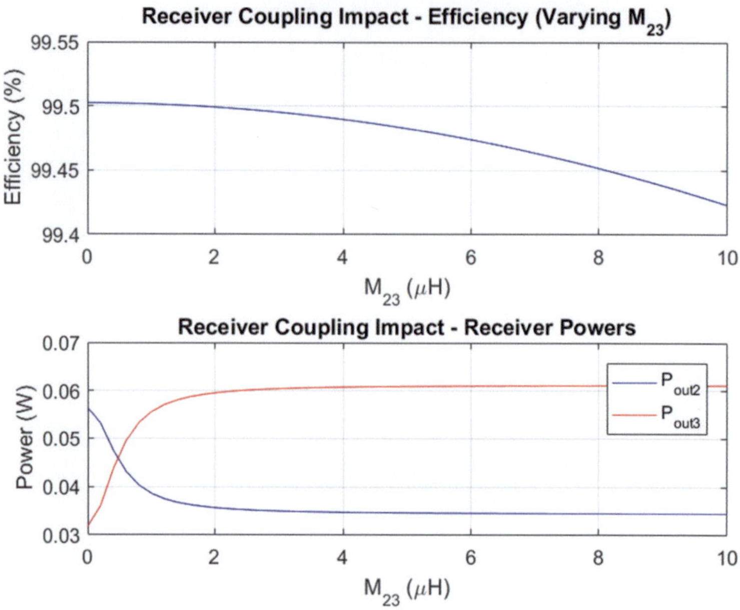

Fig. 8 Power transfer efficiency (PTE) versus receiver coupling impact M_{23}

above 90% across the range, with a peak at 20 μH (nominal alignment). The robustness is due to the iterative optimization of $C_{1\text{match}}$, which compensates for coupling variations. This result is critical for practical WPT systems, where misalignment is common in real-world scenarios.

Frequency Splitting Compensation

In strong coupling scenarios ($M_{12} = 30\,\mu$H, $M_{13} = 25\,\mu$H, $M_{23} = 10\,\mu$H), frequency splitting can degrade PTE. Figure 10 shows PTE versus frequency (6.9 MHz), with adaptive matching maintaining a single efficiency peak at 6.78 MHz (≈95%). This indicates effective compensation for splitting effects, ensuring stable performance in high-coupling conditions, a critical feature for closely spaced receivers.

4 Comparison with Traditional Methods

To highlight the advantages of the adaptive approach, PTE is compared for AI-driven versus static matching as R_{load2} varies from 5 to 50 Ω. Figure 11 shows that AI-driven matching achieves up to 15% higher PTE than static matching, particularly at non-nominal loads. This improvement is due to the dynamic optimization of $C_{1\text{match}}$, underscoring the superiority of adaptive systems over traditional fixed-tuning methods in dynamic environments (Fig. 12).

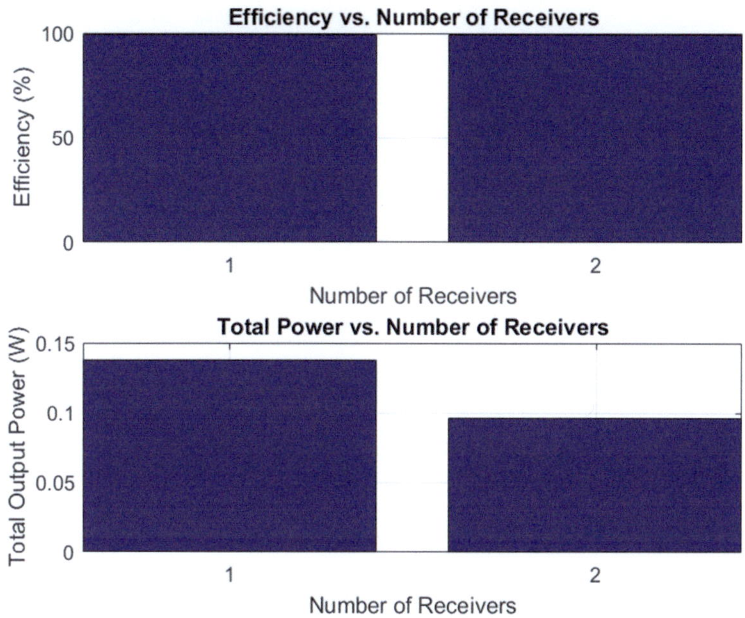

Fig. 9 Power transfer efficiency (PTE) and output power versus number of receivers (1 and 2)

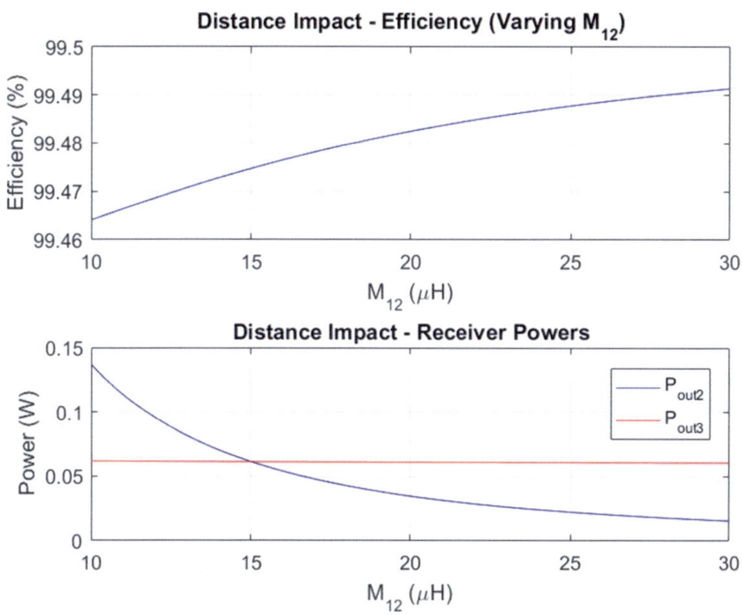

Fig. 10 PTE versus Coil misalignment, modeled by varying M_{12}

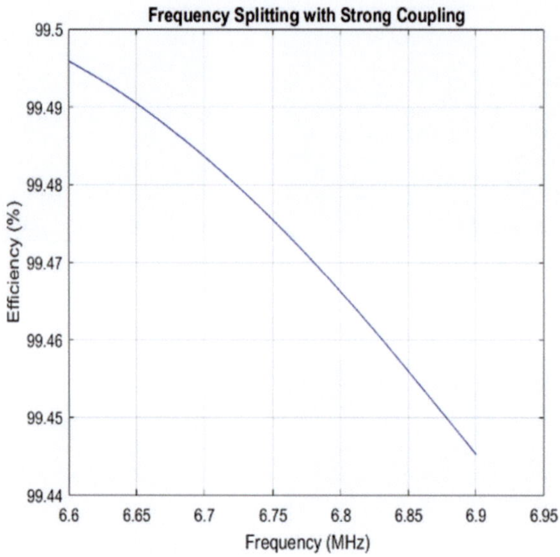

Fig. 11 Power transfer efficiency (PTE) versus frequency under strong coupling

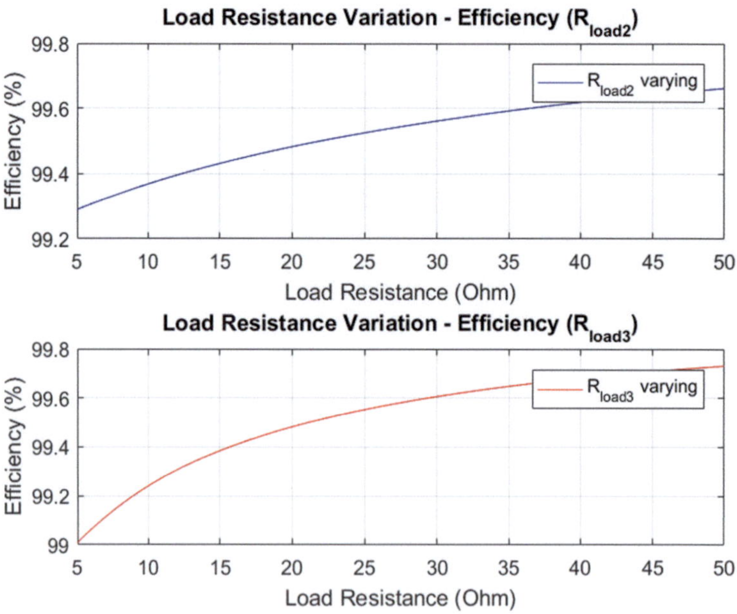

Fig. 12 PTE versus R_{load2} for AI-driven and static matching

5 Discussion

The simulation results demonstrate that the proposed adaptive impedance matching network, using iterative optimization as a proxy for AI-driven control, significantly enhances the performance of multi-receiver WPT systems at 6.78 MHz in practice, C_{1match} can be tuned by lightweight AI policies (e.g., supervised regression for load estimation or a bandit-style optimizer for online PTE maximization), maintaining responsiveness under time-varying coupling and loads.

The high PTE (95% at resonance) and robust load balancing (power difference < 15%) validate the system's ability to handle dynamic load variations. The mitigation of cross- coupling and frequency splitting effects further ensure stable performance in complex scenarios. While the simulated AI approach (iterative optimization) is effective, implementing true AI (e.g., neural networks or reinforcement learning) could further improve responsiveness and accuracy, particularly for real-time load estimation and scalability to more receivers. The low power consumption of the control system (60 mW) supports its feasibility for practical deployment. However, the lack of time-domain analysis limits insights into transient behavior, which could be addressed using Simulink or hardware testing in future work.

A comparison table has been added to compare the results of adaptive based impedance matching verses non-adaptive based impedance matching in multi receiver wireless power transfer. The Table 1 showed that adaptive based impedance matching approach demonstrate significant improvements in power transfer efficiency. This improvement is due to the dynamic optimization of C1match, underscoring the superiority of adaptive systems over traditional fixed-tuning methods in dynamic environments.

Table 1 Performance comparison of WPT system operating at 6.78 MHz with multiple receivers

WPT method	Resonant frequency (MHz)	Multi-receiver transmission efficiency (%)	References
Fixed impedance matching	6.78	78	[16]
Relay coil system	6.78	70–85	[17]
Planar Tx coil with switchable IMNs	6.78	93	[18]
Charging bowl with multiple Rx coils and controllable circuits	6.78	68–80	[19]
Printed spiral half-cube shaped Tx structure	6.78	96.1	[1]
Adaptive impedance matching	**6.78**	**95**	**This work**

6 Conclusion

This study presents a novel adaptive impedance matching network for dynamic load balancing in multi-receiver WPT systems operating at 6.78 MHz, with a simulated AI-driven approach using iterative optimization. The results demonstrate significant improvements in power transfer efficiency (up to 95%), load estimation accuracy (<10% error), and power distribution balance (<15% difference), while effectively mitigating cross-coupling and frequency splitting effects. The systems robustness to misalignment and scalability to multiple receivers, combined with low control power consumption (60 mW), highlights its potential for practical WPT applications, such as wireless charging of multiple devices. Compared to traditional static matching, the adaptive approach achieves up to 15% higher PTE under varying load conditions. Future work should focus on implementing real AI algorithms (e.g., neural networks or reinforcement learning) using modern MATLAB toolboxes or embedded systems, conducting time-domain analysis, and developing a hardware prototype to validate these findings in real-world scenarios. These advancements could further enhance the efficiency and applicability of multi-receiver WPT systems in dynamic environments.

References

1. Naveed I, Tan L, Yameen M, Hussain A (2025) Free-positioning wireless power transfer for multiple receivers using a half-cube shaped transmitting coil. Wirel Power Transf 12:e021. https://doi.org/10.48130/wpt-0025-0016
2. Clerckx B, Zhang R, Schober R, Ng DWK, Kim DI, Poor HV (2019) Fundamentals of wireless information and power transfer: from RF energy harvester models to signal and system designs. IEEE J Sel Areas Commun 37(1):4–33
3. Ahn D, Kim S (2016) Effect of coupling between multiple transmitters or receivers on wireless power transfer. IEEE Trans Ind Electron 63(3):1504–1513
4. Zhang Y, Zhao Z, Chen K (2018) Frequency-splitting analysis of wireless power transfer systems with multiple receivers. IEEE Trans Power Electron 33(7):6140–6151
5. Li X, Zhang C, Wang Z (2019) Adaptive impedance matching for wireless power transfer using machine learning. IEEE Trans Ind Electron 66(11):8536–8545
6. Wang Z-H, Chen Y-P, Li Y (2017) Load detection and power flow control for multi-receiver wireless power transfer systems. IEEE Trans Power Electron 32(7):5129–5139
7. Naveed I, Zhao G, Yamin Z, Gul W (2020) Steady state performance analysis of DFIG with different magnetizing strategies in a pitch-regulated variable speed wind turbine. In: 2020 4th international conference on power and energy engineering (ICPEE). IEEE, pp 174–179. https://doi.org/10.1109/ICPEE51316.2020.9310982
8. Liu Z, Wang Y, Guo Y, Ren G (2020) Multi-receiver magnetic resonance wireless power transfer with dynamic tuning. IEEE Trans Power Electron 35(8):8671–8682
9. Yameen MZ, Junejo AK, Lu Z, Siddiqui RA, El-Sousy FF, Naveed I (2025) Hybrid GOA and PSO optimization for load frequency control in renewable multi-source dual area power systems. Sci Rep 15(1):17549. https://doi.org/10.1038/s41598-025-01481-6
10. Chen H, Sun M, Ding X, Cai Z (2021) High-density multi-coil wireless power transfer system design. IEEE Trans Electromagn Compat 63(2):450–459
11. Lee S, Park J, Lee H (2021) AI-driven control of resonant frequency in wireless power transfer. IEEE Trans Ind Electron 68(12):12234–12242

12. Xu J, Ping H, Liu Z (2022) Adaptive coil alignment compensation in wireless power transfer using feedback. IEEE Trans Magn 58(4):2300210
13. Zhao W, Liu Y, Zhang T (2023) Robust control of multi-receiver wireless power transfer under misalignment conditions. IEEE Trans Control Syst Technol 31(1):295–305
14. He K, Ma J (2023) Reconfigurable impedance matching network for dynamic wireless power transfer. IEEE Trans Power Electron 38(4):3456–3465
15. Singh R, Bhattacharya S (2024) Energy efficiency optimization in multi-device wireless charging pads. IEEE J Emerg Sel Top Power Electron 12(4):3890–3899
16. Kim J, Jeong J (2014) Range-adaptive wireless power transfer using multiloop and multiturn receiver coils. IEEE Trans Antennas Propag 62(12):6494–6503. https://doi.org/10.1109/TAP.2014.2361131
17. Liu F, Chen K, Zhao Z, Li K, Yuan X (2018) Transmitter-side control of both the CC and CV modes for the wireless EV charging system with weak communication. IEEE Trans Power Electron 33(9):7112–7121. https://doi.org/10.1109/TPEL.2017.2772823
18. Kim J, Kim D, Park Y (2016) Free-positioning wireless power transfer to multiple devices using a planar transmitting coil and switchable impedance matching networks. IEEE Trans Microw Theory Tech 64(11):3714–3722
19. Feng J, Li Q, Lee FC, Fu M (2019) Transmitter coils design for free-positioning omnidirectional wireless power transfer system. IEEE Trans. Ind. Inform. 15(8):4656–4664. https://doi.org/10.1109/TII.2019.29082

Hybrid Single-Dual Mode Overmodulation Strategy for Electrolytic Capacitorless Permanent Magnet Synchronous Motor Drives

Cong Bai, Bo Wang, Zhonggang Yin, Yuehan Li, and Liangliang Wang

Abstract Permanent magnet synchronous motor (PMSM) drive systems that utilize small-capacity film capacitors in DC bus instead of electrolytic capacitors have advantages such as long lifespan, compact size, and low cost. However, when traditional dual-mode overmodulation strategies are employed to enhance DC bus voltage utilization, significant fluctuations in the voltage modulation index can cause the output voltage vector to frequently enter and exit the square-wave modulation state. This results in substantial distortion of the output voltage. On the other hand, conventional single-mode overmodulation strategies, although producing less output voltage distortion, are unable to achieve the square-wave modulation state, leading to limited output voltage capability. To balance output voltage capability and distortion, this paper investigates a hybrid single-dual mode overmodulation strategy for electrolytic capacitorless PMSMs. By incorporating a hybrid judgment step after the condition that determines whether the PMSM enters square-wave modulation in the traditional dual-mode overmodulation strategy, the allocation ratio of conventional single and dual-mode overmodulation strategies within the sectors is adjusted, thereby improving the control performance of PMSM drives in the overmodulation region. Simulation results validate the effectiveness of the proposed method.

Keywords Permanent magnet synchronous motor · Electrolytic capacitorless · Hybrid single-dual mode · Overmodulation strategy

1 Introduction

Conventional permanent magnet synchronous motor (PMSM) drive systems generally require large-capacitance electrolytic capacitors on the DC bus for energy storage, filtering, and voltage stabilization [1–3]. However, electrolytic capacitors have a short lifespan and cause environmental pollution. Existing improvements

C. Bai · B. Wang · Z. Yin (✉) · Y. Li · L. Wang
Xi'an University of Technology, Xi'an, Shaanxi, China
e-mail: zhgyin@xaut.edu.cn

G. Mu et al. (eds.), *Proceedings of the 2nd Electrical Artificial Intelligence Conference, Volume 1*, Lecture Notes in Electrical Engineering 1587,
https://doi.org/10.1007/978-981-95-7932-7_19

primarily involve replacing them with small-capacity film capacitors, which offer advantages such as longer lifespan, smaller size, and lower cost. This approach shows broad applications in fields such as heating, ventilation, and air conditioning [4, 5].

The overmodulation strategy can increase the fundamental amplitude of the inverter output voltage by up to 10.27%, thereby effectively enhancing the motor's speed and torque, which holds broad application prospects in fields such as electric vehicles and electric locomotives [6]. Currently, based on the division of the overmodulation region, overmodulation strategies can be classified into single-mode and dual-mode strategies. Single-mode overmodulation strategy applies one adjustment strategy across the entire overmodulation region. A commonly used single-mode strategy is the minimum phase error method. When the amplitude of the reference voltage vector exceeds the voltage boundary, this strategy scales the reference voltage vector proportionally to constrain the output voltage vector within the regular hexagon voltage boundary. This method is relatively simple to implement, resulting in low voltage harmonic content, and is currently the most widely used single-mode overmodulation strategy. Dual-mode overmodulation strategy divides the entire overmodulation region into two areas, applying different adjustment strategies to each one [7, 8]. On the basis of calculating the reference angle and holding angle, [7] used voltage extension to extend the reference voltage vector from the inscribed circle to the voltage boundary, and then from the voltage boundary to the square wave modulation state. On this basis, [8] conducted segmented linear optimization of the calculation of reference angle and holding angle based on the reference voltage modulation coefficient, greatly reducing the computational complexity and promoting engineering applications.

The use of small-capacity film capacitor on the DC bus side leads to a 6 k-times grid frequency ripple in the DC bus output voltage by the rectifier bridge, resulting in significant fluctuations in the voltage modulation index [9–12]. When the traditional dual-mode overmodulation strategy is employed to improve DC bus voltage utilization, the substantial variation in the modulation index causes the output voltage vector to frequently enter and exit the square-wave modulation state, leading to considerable output voltage distortion [13]. Although the traditional single-mode overmodulation strategy has a lower output voltage distortion, it cannot enter the square wave modulation state and has a lower ability to output voltage [14].

To balance output voltage capability and distortion degree, this paper investigates a hybrid single-dual mode overmodulation strategy for electrolytic capacitorless PMSMs. The study begins with an analysis of the electrolytic capacitorless PMSM system, including its structural analysis and harmonic analysis. Subsequently, the implementation methods of traditional single and dual-mode overmodulation strategies are introduced, and the output voltage distortion and voltage capabilities in the electrolytic capacitorless PMSM system are analyzed, respectively. To achieve a trade-off between output voltage capability and distortion, a hybrid judgment step is incorporated after the condition determining whether the motor enters the square-wave modulation state in the traditional dual-mode overmodulation strategy.

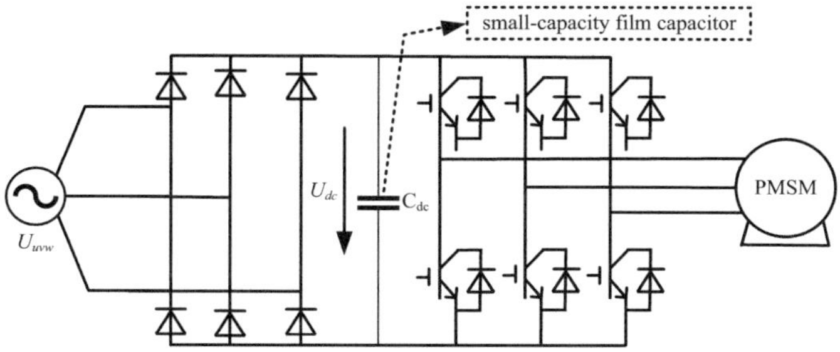

Fig. 1 Structure of electrolytic capacitorless PMSM drive system

This adjustment regulates the distribution ratio of the conventional single and dual-mode overmodulation strategies within the sectors, thereby enhancing the motor's control performance in the overmodulation region. Simulation results validate the effectiveness of the proposed method.

2 Analysis of Electrolytic Capacitorless PMSM Drive System

2.1 Structural Analysis of Electrolytic Capacitorless PMSM System

The electrolytic capacitorless PMSM drive system mainly consists of a three-phase diode uncontrolled rectifier bridge, a small-capacity film capacitor, a three-phase voltage-source two-level inverter, and a permanent magnet synchronous motor. The system structure is shown in Fig. 1.

2.2 Harmonic Analysis of Electrolytic Capacitorless PMSM System

Due to the use of small-capacity film capacitor on the inverter DC side, the limited energy stored in the capacitor leads to a $6k$ grid frequency ripple in the DC side voltage. The DC bus voltage U_{dc} can be expressed as,

$$U_{\text{dc}} = U_{\text{dc},0} + \sum_{\text{k}=1}^{\infty} U_{\text{dc,k}} \sin(6k\omega_{\text{g}}t + \varphi_{\text{g}}) \tag{1}$$

where $U_{dc,0}$ is the average value of the DC bus voltage, $U_{dc,k}$ and φ_k are the amplitude and phase of the kth harmonic, and ω_g is the angular frequency of grid voltage.

The maximum value U_{dcmax} and minimum value U_{dcmin} of U_{dc} are,

$$\begin{cases} U_{dcmax} = \sqrt{6} * U_{phase} \\ U_{dcmin} = \frac{\sqrt{3}}{2} * U_{dcmax} \end{cases} \tag{2}$$

where U_{phase} is the RMS value of the grid-side phase voltage.

Due to the significant periodic fluctuation of the DC bus voltage, the same frequency harmonics will appear in the motor voltage. Therefore, when only considering the disturbance caused by the DC bus voltage fluctuation, the reference voltage modulation index m and output voltage modulation index MI are obtained as,

$$\mathrm{m} = \frac{3U_{ref}}{2U_{dc}} = \mathrm{M}_0 + \sum_{\mathrm{k}=1}^{\infty} \mathrm{M}_{\mathrm{k}} \sin(6k\omega_g + \varphi_{gr}) \tag{3}$$

$$\mathrm{MI} = \frac{\pi U_o}{2U_{dc}} = \mathrm{MI}_0 + \sum_{k=1}^{\infty} \mathrm{MI}_{\mathrm{k}} \sin(6k\omega_g + \varphi_{go}) \tag{4}$$

where U_{ref} and U_o are the magnitudes of the reference voltage vector and output voltage vector, respectively. M_0 and MI_0 are the average values of the reference voltage modulation index and output voltage modulation index, respectively. M_{k}, φ_{gr} and MI_{k}, φ_{go} are the magnitudes and phases of the kth harmonic of the reference voltage and output voltage modulation indices, respectively.

3 Analysis of Traditional Overmodulation Strategies

3.1 Analysis of Traditional Dual-Mode Overmodulation Strategy

Taking the first sector as an example, the voltage adjustment strategy of the traditional dual-mode overmodulation strategy is shown in Fig. 2. $\boldsymbol{U}_4$ and $\boldsymbol{U}_6$ are the adjacent voltage space vectors in the first sector, $\boldsymbol{U}_{ref1}$ and $\boldsymbol{U}_{ref2}$ are the reference voltage vectors, and $\boldsymbol{U}_{o1}$ and $\boldsymbol{U}_{o2}$ are the output voltage vectors.

(1) When the reference voltage vector is within the regular hexagon voltage boundary,

$$\begin{cases} T_4 = T_{4\_ref} \\ T_6 = T_{6\_ref} \end{cases} \tag{5}$$

Fig. 2 Diagram of voltage regulation of traditional dual-mode overmodulation strategy

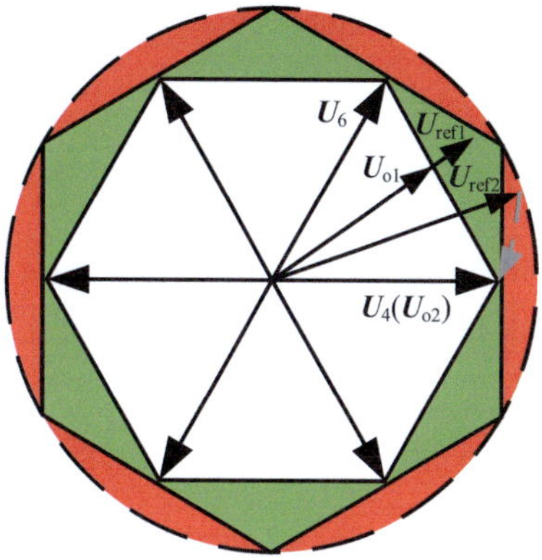

where T_{4_ref} and T_{6_ref} are the reference action times of the adjacent voltage space vectors in the first sector, respectively. T_s is the inverter switching period. T_4 and T_6 are the actual action times of adjacent voltage space vectors in first sector, respectively.

(2) When the reference voltage vector $\boldsymbol{U}_{ref1}$ exceeds the regular hexagon voltage boundary but does not exceed the circumscribed regular hexagon voltage boundary ($1 > m \geq 0.866$, $0.9517 > MI \geq 0.9069$), the phase of the reference voltage vector is kept unchanged, and it is proportionally scaled and constrained to the regular hexagon voltage boundary.Thus,

$$\begin{cases} T_4 = \dfrac{T_{4\_ref}}{T_{4\_ref}+T_{6\_ref}} T_s \\ T_6 = \dfrac{T_{6\_ref}}{T_{4\_ref}+T_{6\_ref}} T_s \end{cases} \tag{6}$$

(3) When the reference voltage vector $\boldsymbol{U}_{ref2}$ exceeds the circumscribed regular hexagon voltage boundary ($1.2 \geq m \geq 1$, $1 \geq MI \geq 0.9517$), the motor enters the square-wave modulation state. Thus,

$$\begin{cases} T_4 = T_s \\ T_6 = 0 \end{cases}, \quad T_{4\_ref} > T_{6\_ref} \tag{7}$$

$$\begin{cases} T_4 = 0 \\ T_6 = T_s \end{cases}, \quad T_{6\_ref} >= T_{4\_ref} \tag{8}$$

The relationship between output voltage and reference voltage modulation index MI and m in traditional dual-mode overmodulation strategy is shown in Fig. 3.

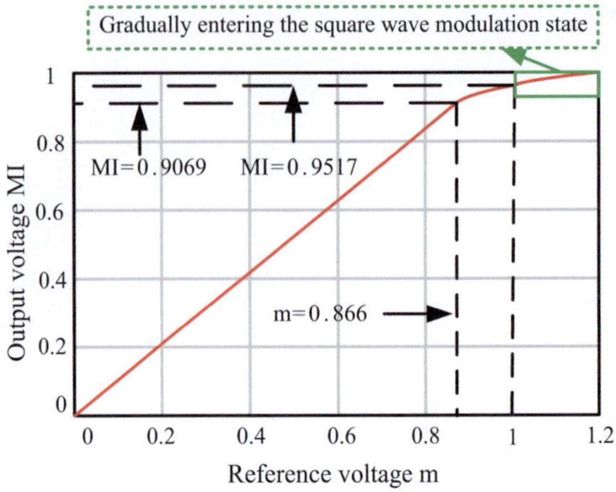

Fig. 3 Relationship between MI and m of the traditional dual-mode overmodulation strategy

3.2 *Analysis of Traditional Single-Mode Overmodulation Strategy*

Taking the first sector as an example, the voltage adjustment strategy of the traditional single-mode overmodulation strategy is shown in Fig. 4. $\boldsymbol{U}_4$ and $\boldsymbol{U}_6$ are the adjacent voltage space vectors, $\boldsymbol{U}_{\text{ref3}}$ is the reference voltage vector, and $\boldsymbol{U}_{\text{o3}}$ is the output voltage vector.

(1) When the reference voltage vector is within the regular hexagon voltage boundary ($0.866 > \text{m} \geq 0$, $0.9069 > \text{MI} \geq 0$),

$$\begin{cases} T_4 = T_{4\_\text{ref}} \\ T_6 = T_{6\_\text{ref}} \end{cases} \tag{9}$$

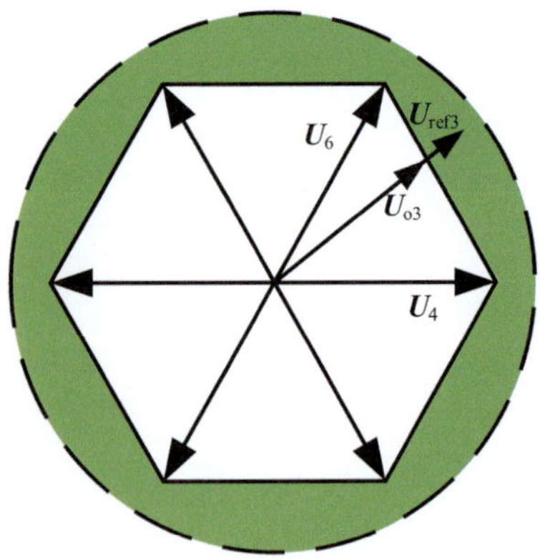

Fig. 4 Diagram of voltage adjustment for traditional single-mode overmodulation strategy

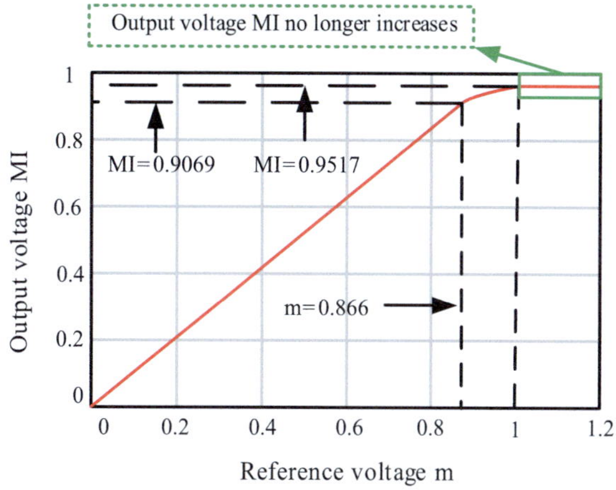

Fig. 5 Relationship between MI and m of traditional single-mode overmodulation strategy

(2) When the reference voltage vector $\boldsymbol{U}_{\mathrm{ref3}}$ exceeds the regular hexagon voltage boundary ($1.2 \geq \mathrm{m} \geq 0.866$, $0.9517 \geq \mathrm{MI} \geq 0.9069$), the phase of the reference voltage is kept unchanged, and it is proportionally scaled and constrained to the regular hexagon voltage boundary. Thus,

$$\begin{cases} T_4 = \dfrac{T_{4\_\mathrm{ref}}}{T_{4\_\mathrm{ref}}+T_{6\_\mathrm{ref}}} T_s \\ T_6 = \dfrac{T_{6\_\mathrm{ref}}}{T_{4\_\mathrm{ref}}+T_{6\_\mathrm{ref}}} T_s \end{cases} \tag{10}$$

The relationship between output voltage and reference voltage modulation indices MI and m of traditional single-mode overmodulation strategy is shown in Fig. 5.

3.3 Output Voltage Performance Analysis of Two Overmodulation Strategies

Due to significant periodic fluctuations in the DC bus voltage, there is substantial variations of the reference voltage modulation index m. Especially when the reference voltage vector reaches the vertices of the circumscribed regular hexagon voltage boundary, the motor may uncontrollably enter or exit the six-step operation state, causing significant sudden jumps and retractions in the phase of the output voltage vector. The trajectory of the reference voltage modulation index of the traditional dual-mode overmodulation strategy is illustrated in Fig. 6, and the output α-axis voltage u_{α} for the traditional dual-mode overmodulation strategy is shown in Fig. 7.

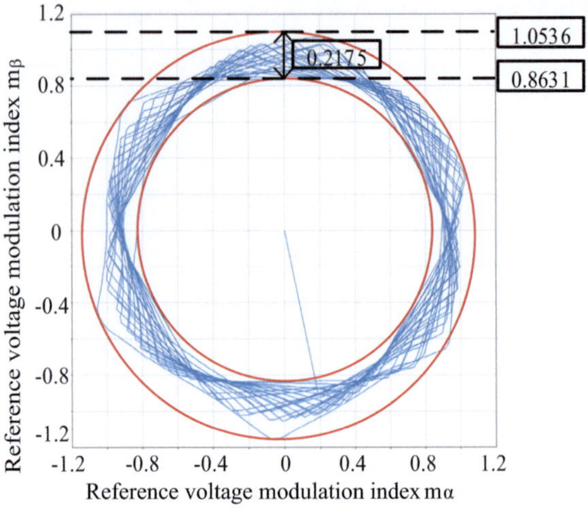

Fig.6 Trajectory of reference voltage modulation index m of the traditional dual-mode overmodulation strategy

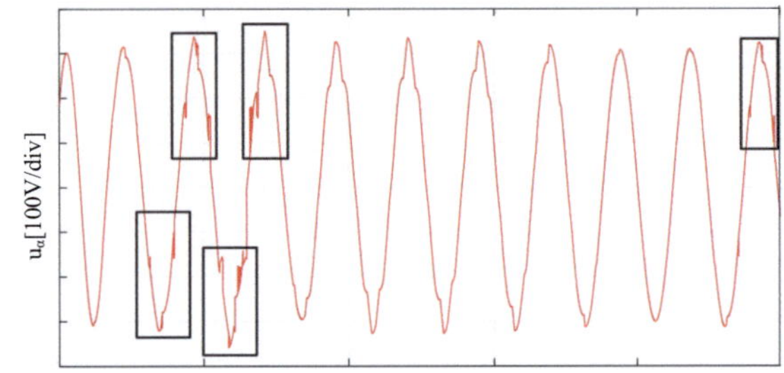

Fig. 7 u_α of traditional dual-mode overmodulation strategy

The trajectory of the reference voltage modulation index m and the output α-axis voltage u_α of the traditional single-mode overmodulation strategy are shown in Figs. 8 and 9, respectively.

By combining Figs. 3, 6, and 7, it is evident that when the traditional dual-mode overmodulation strategy is adopted, the motor's operating state in the modulation region becomes highly chaotic. It continuously enters and exits the square-wave modulation mode, and may even revert to the linear modulation region, resulting in significant distortion in both the phase and amplitude of the output voltage, thereby degrading the motor's output performance.

By combining Figs. 5, 8 and 9, it is clear that when the traditional single-mode overmodulation strategy is adopted, although the output voltage distortion is relatively smaller compared to the traditional dual-mode overmodulation strategy due to its inherent inability to enter the square-wave modulation mode, the output voltage capability is also limited.

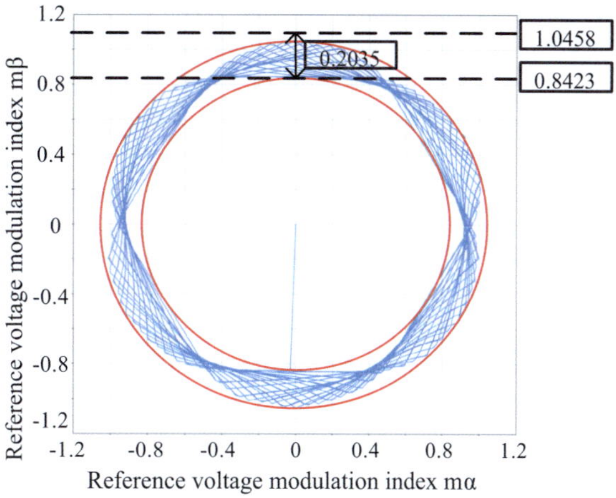

Fig. 8 Trajectory of reference voltage modulation index m of the traditional single-mode overmodulation strategy

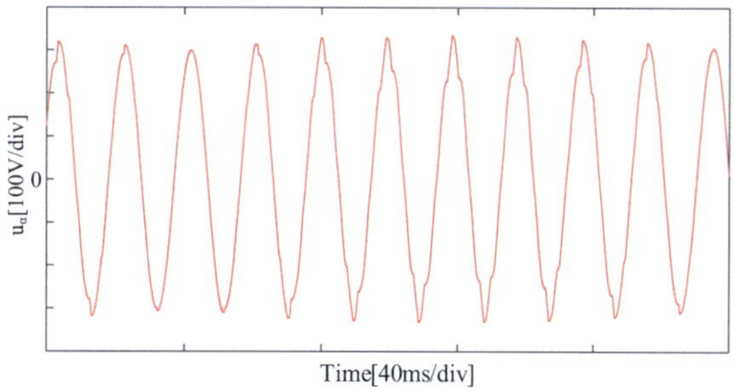

Fig. 9 u_{α} of traditional single-mode overmodulation strategy

4 Hybrid Single-Dual Mode Overmodulation Strategy for Electrolytic Capacitorless PMSM System

On the basis of the previous section, neither the traditional dual-mode strategy nor the single-mode overmodulation strategy can effectively balance the output voltage capability and distortion degree. Therefore, this section investigates a hybrid single-dual mode overmodulation strategy.

To adjust the allocation ratio of the traditional single and dual-mode overmodulation strategies within the sector, an allocation ratio coefficient DR (DR $\geq$ 1) is defined. When DR $=$ 1, this strategy is equivalent to the traditional dual-mode overmodulation strategy. When DR $>$ 1, the proportion of the single-mode overmodulation strategy gradually increases.

Taking the first sector as an example, the voltage adjustment strategy of the hybrid single-dual mode overmodulation strategy is shown in Fig. 10. $\boldsymbol{U}_{\text{ref4}}$ and $\boldsymbol{U}_{\text{ref5}}$ are the reference voltage vectors, and $\boldsymbol{U}_{\text{o4}}$ and $\boldsymbol{U}_{\text{o5}}$ are the output voltage vectors.

Fig. 10 Diagram of voltage regulation of hybrid single-dual mode overmodulation strategy

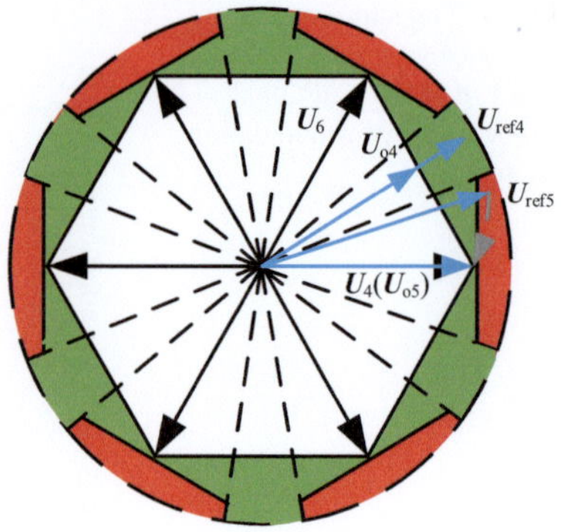

(1) When the reference voltage vector is within the regular hexagon voltage boundary (0.866 > m ≥ 0, 0.9069 > MI ≥ 0),

$$\begin{cases} T_4 = T_{4\_ref} \\ T_6 = T_{6\_ref} \end{cases} \tag{11}$$

(2) When the reference voltage vector $\boldsymbol{U}_{ref4}$ is in the green region shown in Fig. 10, that is, $((T_{4\_ref}+T_{6\_ref}) > T_s)\&((T_{4\_ref}/T_{6\_ref} \le \mathrm{DR})|((T_{4\_ref} + T_{6\_ref}/2) \le T_s))$, or $((T_{4\_ref}+T_{6\_ref}) > T_s)\&((T_{6\_ref}/T_{4\_ref} \le \mathrm{DR})|((T_{4\_ref}/2 + T_{6\_ref}) \le T_s))$, keeping the phase of the reference voltage vector unchanged, perform proportional scaling constraints on it to the regular hexagonal voltage boundary. Thus,

$$\begin{cases} T_4 = \dfrac{T_{4\_ref}}{T_{4\_ref}+T_{6\_ref}} T_s \\ T_6 = \dfrac{T_{6\_ref}}{T_{4\_ref}+T_{6\_ref}} T_s \end{cases} \tag{12}$$

(3) When the reference voltage vector $\boldsymbol{U}_{ref5}$ is in the red area in Fig. 10, that is $(T_{4\_ref}/T_{6\_ref}>\mathrm{DR})\&((T_{4\_ref}+T_{6\_ref}/2)>T_s)$ or $(T_{6\_ref}/T_{4\_ref}>\mathrm{DR})\&((T_{4\_ref}/2+T_{6\_ref}) > T_s)$, the motor begins to enter the square wave modulation state,

$$\begin{cases} T_4 = T_s \\ T_6 = 0 \end{cases}, \quad T_{4\_ref} > T_{6\_ref} \tag{13}$$

$$\begin{cases} T_4 = 0 \\ T_6 = T_s \end{cases}, \quad T_{6\_ref} >= T_{4\_ref}. \tag{14}$$

5 Simulation Results

To verify the effectiveness of the electrolytic capacitorless permanent magnet synchronous motor based on the hybrid single-dual mode overmodulation strategy proposed in this paper, a simulation model was built in the Matlab 2023b/Simulink environment for validation. Figure 11 shows the flowchat of the hybrid single-dual mode overmodulation strategy. Table 1 shows the parameters of the electrolytic capacitorless PMSM drive system used in the simulation. The simulation employs the variable-step ode45 algorithm.

Figure 12 shows the simulation waveforms under steady-state conditions when using the traditional dual-mode overmodulation strategy, with the PMSM driving

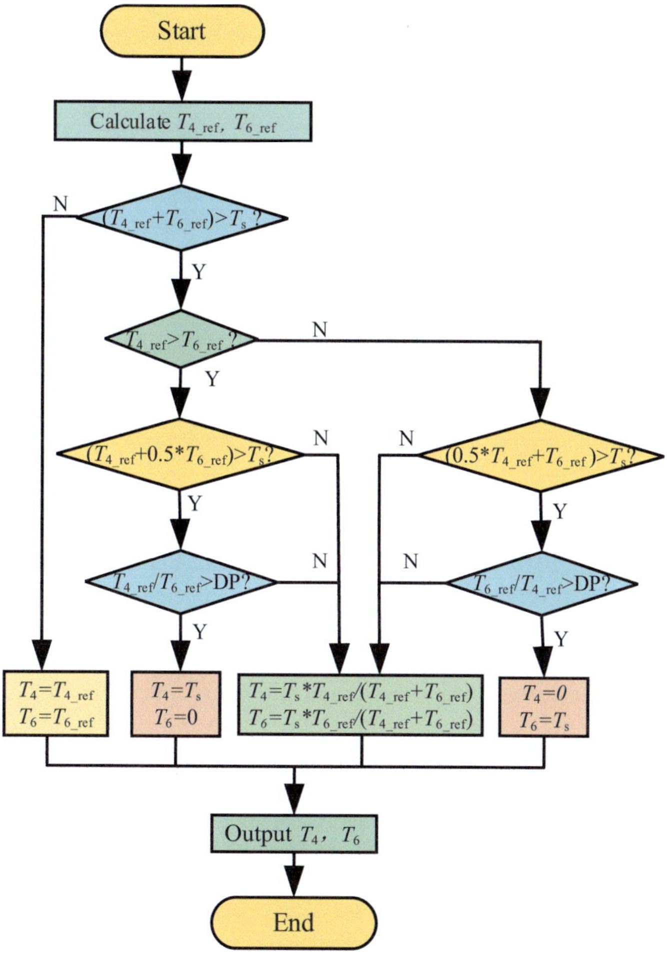

Fig.11 Flowchart of the hybrid single-dual mode overmodulation strategy

Table 1 Parameters of electrolytic capacitorless PMSM system

Parameter	Value	Parameter	Value
DC-link capacitance	50 μF	Stator resistance	2.75 Ω
Rated current	5.8 A	d-axis inductance	35 mH
Rated speed	1000 r/min	q-axis inductance	54 mH
Rated frequency	50 Hz	Number of pole pairs	3
Switching frequency	10 kHz	Rotor flux linkage	0.86 Wb

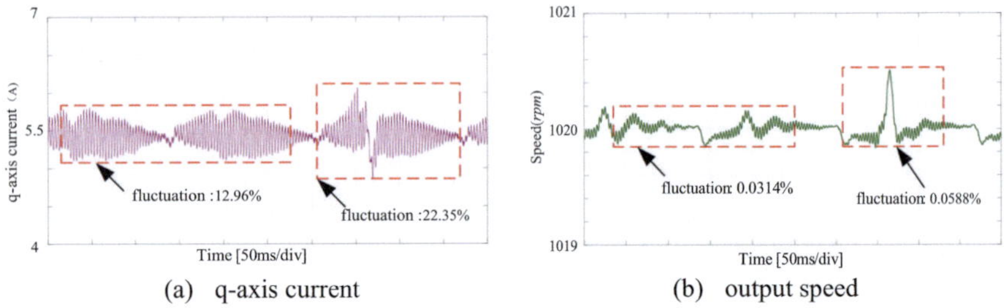

Fig. 12 Simulation waveform with traditional dual-mode overmodulation strategy at 51 Hz

a rated load at 51 Hz (1020 r/min). Figure 12a and b represent the output q-axis current i_q and the output speed n, respectively. It can be observed that the fluctuation of the output q-axis current i_q jumps from 12.96 to 22.35%, and the fluctuation of the output speed jumps from 0.0314 to 0.0588%. When using the traditional dual-mode overmodulation strategy, the motor uncontrollably enters the square-wave modulation mode, leading to sudden changes in the output waveforms and a degradation in control performance.

Figure 13 shows the simulation waveforms under steady-state conditions when using the traditional single-mode overmodulation strategy, with the PMSM driving a rated load at 51 Hz (1020 r/min). Figure 13a and b represent the output q-axis current i_q and the output speed n, respectively. The results indicate that the fluctuation of the output q-axis current i_q is 12.09%, and the fluctuation of the output speed n is 0.0234%. When using the traditional single-mode overmodulation strategy, since the PMSM does not enter the square-wave modulation mode, the issues associated with the traditional dual-mode overmodulation strategy are avoided. Therefore, the control performance of the PMSM is better compared to that of the traditional dual-mode overmodulation strategy.

Figure 14 presents the simulation waveforms under steady-state conditions when employing the hybrid single-dual mode overmodulation strategy, with the motor driving a rated load at 51 Hz (1040 r/min). Figure 14a and b represent the output q-axis current i_q, and the output speed n, respectively. The results show that the fluctuation of the output q-axis current i_q is 12.31%, and the fluctuation of the output speed is 0.0265%. When using the hybrid single-dual mode overmodulation strategy, the

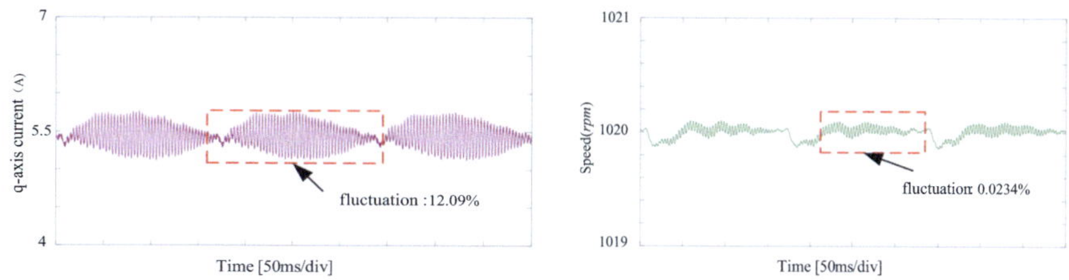

(a) q-axis current (b) output speed

Fig. 13 Simulation waveform with traditional single-mode overmodulation strategy at 51 Hz

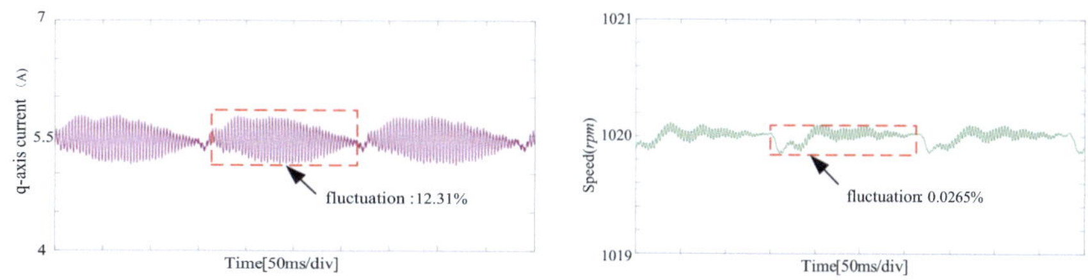

(a) q-axis current (b) output speed

Fig. 14 Simulation waveform with hybrid single-dual mode overmodulation strategy at 51 Hz

motor does not easily enter the square-wave modulation mode. The control performance is close to that of the traditional single-mode overmodulation strategy and superior to the traditional dual-mode overmodulation strategy.

Figure 15 shows the simulation waveforms under steady-state conditions when the traditional dual-mode overmodulation strategy is employed, with the motor driving a rated load at 52 Hz (1040 r/min). Figure 15a and b represent the output q-axis current i_q and the output speed n, respectively. The results indicate that the fluctuation of the output q-axis current i_q is 20.41%, and the fluctuation of the output speed is 0.0727%. When the traditional dual-mode overmodulation strategy is used at a command frequency of 52 Hz, the motor still uncontrollably enters the square-wave modulation mode, leading to significant distortion in the output waveforms and degraded control performance.

Figure 16 shows the output speed waveform under steady-state conditions when the traditional single-mode overmodulation strategy is employed, with the motor driving a rated load at a command frequency of 52 Hz (1040 r/min). The results indicate that when using the traditional single-mode overmodulation strategy, the PMSM fails to reach the command speed due to insufficient output voltage capability. The output speed can only reach a maximum of 1031.5 r/min, exhibiting a significant steady-state error.

Figure 17 presents the simulation waveforms under steady-state conditions when employing the hybrid single-dual mode overmodulation strategy, with the motor driving a rated load at a command frequency of 52 Hz (1020 r/min). Figure 17a and b represent the output q-axis current i_q and the output speed n, respectively.

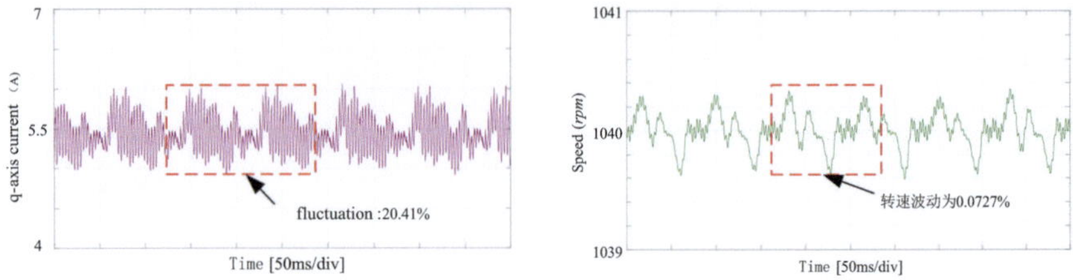

(a) q-axis current (b) output speed

Fig. 15 Simulation waveform with traditional dual-mode overmodulation strategy at 52 Hz

Fig. 16 Simulation waveform of output speed with traditional single-mode overmodulation strategy at 52 Hz

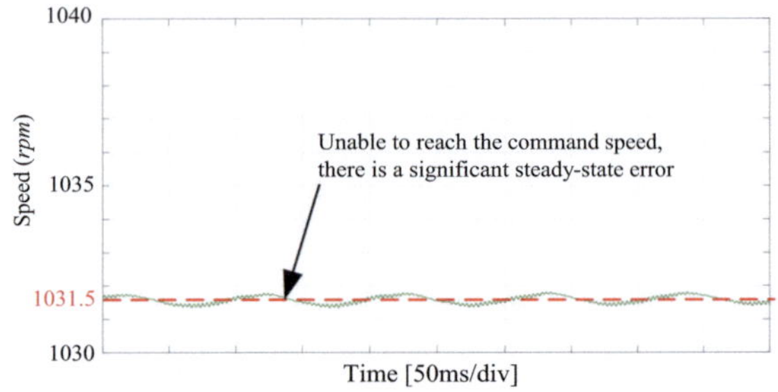

The results indicate that the fluctuation of the output q-axis current i_q is 18.99%, and the fluctuation of the output speed n is 0.0646%. When the hybrid single-dual mode overmodulation strategy is applied at 52 Hz, the motor does not easily enter the square-wave modulation mode and maintains a higher output voltage capability. Therefore, the control performance of PMSM is superior to that of both traditional single-mode and dual-mode overmodulation strategies.

In summary, the hybrid single-dual mode overmodulation strategy of the electrolytic capacitorless PMSMs studied in this paper is superior to the traditional single or dual mode overmodulation strategy. The simulation results have verified the effectiveness of the hybrid single-dual mode overmodulation strategy.

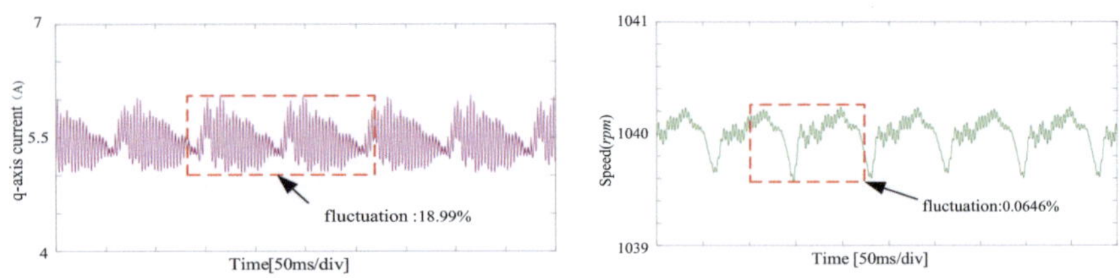

(a) q-axis current (b) output speed

Fig. 17 Simulation waveform with hybrid single-dual mode overmodulation strategy at 52 Hz

6 Conclusion

This paper has investigated a hybrid single-dual mode overmodulation strategy for electrolytic capacitorless PMSMs. To balance the output voltage capability and the distortion level, a hybrid judgment step is incorporated after the stage determining whether the motor enters square-wave modulation in the traditional dual-mode overmodulation strategy. This adjusts the allocation ratio of the traditional single or dual mode overmodulation strategy across sectors, improving the control performance in the overmodulation region. Simulation results demonstrate that the hybrid single-dual mode overmodulation strategy for electrolytic capacitorless PMSMs exhibits superior control performance.

References

1. Du Y , Wang K, Li C (2024) Torque sideband harmonics suppressing in uneven power distribution for N-module in-wheel PMSM drives. IEEE Trans Power Electron 39(2):2020–2034
2. Wang Y, Fang S, Hu J (2023) Active disturbance rejection control based on deep reinforcement learning of PMSM for more electric aircraft. IEEE Trans Power Electron 38(1):406–416
3. Wang S, Zhang G, Wang Q (2023) Torque disturbance compensation method based on adaptive Fourier-Transform for permanent magnet compressor drives. IEEE Trans Power Electron 38(3):3612–3623
4. Li Y, Yin Z, Zhang Y (2025) Overvoltage suppression strategy for electrolytic capacitorless drives using sliding mode controller based on improved variable rate reaching law. IEEE Trans Ind Electron 72(8):7683–7695
5. Ren Z, Ding D, Wang G (2025) Antiovervoltage control strategy for electrolytic capacitorless PMSM drives based on power balance state regulation. IEEE Trans Power Electron 40(6):8005–8015
6. Zhang X, Wang B, Yu Y (2021) Overmodulation index optimization method for torque quality improvement in induction motor field-weakening control. IEEE Trans Ind Electron 68(12):11954–11967
7. Holtz J, Lotzkat W, Khambadkone AM (1993) On continuous control of PWM inverters in the overmodulation range including the sixstep mode. IEEE Trans Power Electron 8(4):546–553
8. Lee D-C, Lee G-M (1998) A novel overmodulation technique for spacevector PWM inverters. IEEE Trans Power Electron 13(6):1144–1151
9. Li Y, Yin Z, Zhang Y (2024) A harmonic suppression strategy based on adaptive synchronous rotating frame transformation for improving the estimation accuracy of sensorless drivers with small capacitors. IEEE Trans Power Electron 39(3):3521–3532
10. Gao R, Ding D, Wang G (2025) Feature extraction of frequency-mapping-based resonance suppression in PMSM drives with low DC-link capacitance and inductance. IEEE Trans Power Electron 40(4):4848–4861
11. Gao R, Ding D, Wang G (2023) Linear modulation region expansion strategy based on voltage vector angle regulation for PMSM drives with small DC-link capacitor. IEEE Trans Power Electron 38(7):8327–8339
12. Ding D, Zhang T, Li B (2025) Adaptive modulation index fluctuation suppression strategy for electrolytic capacitorless PMSM drives. IEEE Trans Trans Electrification 11(2):6406–6417

13. Wang G, Hu H, Ding D (2021) Overmodulation index optimization method for torque quality improvement in induction motor field-weakening control. IEEE Trans Ind Electron 68(12):11954–11967
14. Ding D, Wang G, Zhao N (2019) Enhanced flux-weakening control method for reduced DC-link capacitance IPMSM drives. IEEE Trans Power Electron 34(8):7788–7799

A MFL Mechanism-Guided Intelligent Safety Detection Method for Energy Transmission Systems

Xiangkai Shen, Yifu Ren, He Zhao, Jinxi Gao, Shaofeng Xu, Aning Yang, Jiange Kou, and Zhiguo Yang

Abstract As an important infrastructure for energy transportation, pipelines play a crucial role in ensuring safety and promoting economic development. However, as energy transportation systems remain in service for long periods, various defects may gradually emerge, leading to economic losses or even environmental damage. Recently, object detection methods based on deep learning have achieved promising results in object detection fields. However, due to the lack of mechanistic characteristics and insufficient ability to extract defect features under different operating conditions, deep learning-based defect detection methods still struggle to achieve satisfactory detection results in the defect detection field. To address these issues, this paper proposes an intelligent defect detection method based on the magnetic flux leakage (MFL) mechanism, which effectively combines expert knowledge to enhance the model's detection accuracy. First, a classification regression decoupling (CRD) module is designed to enhance the model's ability to capture discriminative features and spatial information according to the characteristics of different tasks, thereby maintaining high classification accuracy with precise regression positioning. Second, a defect physical characteristics (DPC) loss is designed to effectively guide the model in learning key features related to defect physical characteristics, thereby further improving detection performance.

Keywords Mechanism knowledge · Energy transmission systems · Magnetic flux leakage · Defect detection

---

X. Shen · S. Xu · J. Kou · Z. Yang
Beihang University, Beijing, China

Y. Ren
Tsinghua University, Beijing, China

H. Zhao (✉)
Northeastern University, Shengyang, China
e-mail: z2110304@163.com

J. Gao · A. Yang
Liupanshan Laboratory, Yingchuan, China

G. Mu et al. (eds.), *Proceedings of the 2nd Electrical Artificial Intelligence Conference, Volume 1*, Lecture Notes in Electrical Engineering 1587,
https://doi.org/10.1007/978-981-95-7932-7_20

1 Introduction

With the continuous growth of global demand for oil and natural gas resources, energy transportation systems play an indispensable role in modern society. However, long-term operation under harsh conditions exposes energy transportation systems to multiple risks such as mechanical damage and chemical corrosion, leading to various structural defects that pose serious threats to economic development and human safety. Therefore, conducting regular inspections of energy transportation systems is of great significance. Magnetic flux leakage (MFL) testing, as an important non-destructive testing technique, can effectively reveal potential defect information within energy transportation systems. Researchers can effectively ensure the reliable operation of energy transportation systems by analyzing MFL signals.

Currently, deep learning-based detection methods have been widely applied in the defect detection fields. Deep learning methods can automatically learn the distinguishing features of defects through an end-to-end data-driven approach, effectively improving the efficiency and accuracy of detection.

In [1], an adaptive image segmentation network was designed, which can automatically capture multi-scale contextual information, thereby significantly improving defect detection capabilities in complex scenes. In [2], an enhanced multi-scale feature fusion defect detection method was proposed, which efficiently fuses rich semantic information by introducing hollow space pyramid pooling and feature pyramid networks, thereby improving detection accuracy and network generalization capabilities. In [3], a coarse-to-fine defect detection process was presented, effectively balancing computational efficiency and detection performance. In [4], a convolutional neural network-based defect detection method combining feature enhancement and ring fusion was proposed, which enhances shallow-layer features and fuses information from different levels to effectively improve the detection accuracy of small defects. In [5], a convolutional neural network-based automatic defect signal classification method was designed, which utilizes rectified linear units to enhance the extraction capability of sparse features and strengthen defect detection performance. In [6], a cascaded regional convolutional neural network defect detection method combining feature pyramids and online hard example mining was designed, which significantly improves detection accuracy in complex environments through end-to-end training. In [7], An MFL weak defect detection method based on a two-stage heterogeneous signals mutual supervision method has been proposed. This method effectively improves the detection capability of weak defects by jointly extracting and supervising axial and radial signal features, and combines improved nonmaximum suppression to refine the identification results.

Although the above methods have achieved certain detection effects in defect detection tasks, they are purely data-driven methods and fail to fully explore and utilize the mechanistic knowledge related to defects. This not only reduces the interpretability of the model but also affects the detection accuracy. In addition, the above methods are difficult to simultaneously accommodate the classification and regression tasks in defect detection and fail to extract differentiated features according to the feature requirements of different tasks, thereby limiting the further improvement of overall detection performance.

Based on the above analysis, this paper proposes a MFL mechanism-based intelligent method for MFL defect detection. Overall, the innovations of this paper are as follows:

(1) A classification regression decoupling (CRD) module is designed. This can decouple the classification task and the regression task so that the model can mine more detailed and discriminative features in each task according to the characteristics of the respective tasks, thus enhancing the feature representation of defects and ultimately optimizing the overall defect detection effect.
(2) A defect physical characteristics (DPC) loss is designed, which can inject defect mechanism information into the model training process, effectively guiding the network to focus on key feature expressions related to defects, thereby significantly improving defect detection accuracy.

2 The Proposed Method

2.1 The Architecture of the Proposed Method

Figure 1 shows the overall architecture of the proposed method in this paper. First, data is input into the backbone to extract multi-level feature information. The extracted features are then used as input for subsequent modules. Specifically, a CRD module is designed to enhance the model's ability to extract discriminative features and obtain precise localization information based on the characteristics of different tasks. In addition, a DPC loss is designed to guide the model in effectively learning key features related to defects.

2.2 Classification and Regression Decoupling Module

In previous defect detection methods, it is common to use the same features for both classification and localization tasks. However, classification and localization tasks have different properties, which require the design of specialized model structures for classification and localization tasks to further improve the performance of the models in different tasks. Based on the above analysis, a CRD module is designed, which significantly improves the classification performance and localization accuracy of the model by decoupling and optimizing the classification and localization tasks.

First, a deformable convolution is designed to enhance the network's ability to capture multi-scale information. Specifically, unlike traditional convolution, deformable convolution can flexibly adjust the perceptual field to adapt to changes

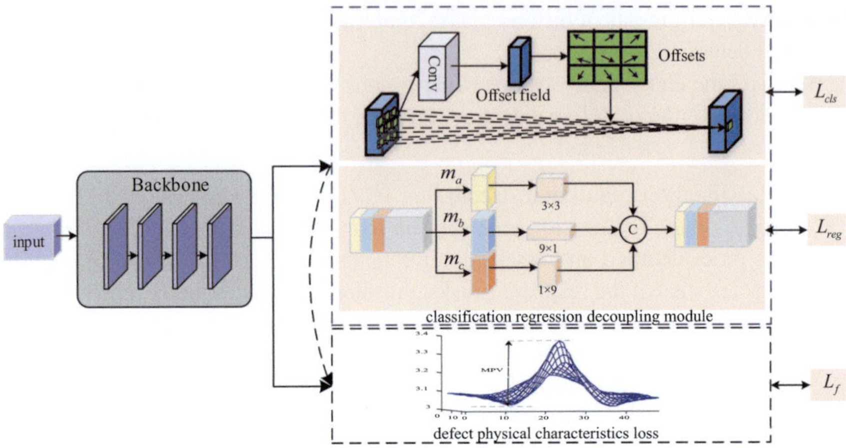

Fig. 1 Architecture of the proposed method

in object shape and scale by learning adaptive sampling locations. This mechanism enables the model to focus more precisely on the key regions, so as to extract more detailed and discriminative features. Therefore, the proposed network provides a more reliable basis for classification tasks, and significantly improves the classification accuracy and robustness. The specific equation is as follows:

$$m(p_0) = \sum_{(p_n \in R)} w(p_n) \cdot z_o(p_0 + \Delta p(p_0)) \tag{1}$$

where p_0 is the target position, p_n is each relative position on the convolution kernel, R is the convolution kernel range, and $\Delta p(p_0)$ is the learned offset.

Second, considering the diversity of defect types in industrial data and the possible defect types with large length and width, a grouped convolution is designed to capture local information, multi-scale long bar features, and cross-direction large-scale features in the feature map, respectively. Specifically, the input features m are evenly divided into four feature maps m_a, m_b, and m_c. The feature maps m_a, m_b, and m_c are processed by 3×3, 9×1, and 1×9 convolution operations to extract local details, multi-scale features, and directional information, as follows.

$$m_f = \mathbb{C}((W_{3\times 3} \otimes m_a + b), (W_{9\times 1} \otimes m_b + b), (W_{1\times 9} \otimes m_c + b)) \tag{2}$$

where m_f is the feature map after concatenating. Local information extraction helps to capture fine target details, multi-scale long bar features can adapt to changes in target shape and orientation, and cross-direction large-scale features provide rich global contextual information. Therefore, the model is able to effectively capture long-range dependencies while focusing on local details, which can significantly enhance the expressiveness and adaptability of features.

Then, the m_d concatenates with the m_f, which enhances the feature expression ability and retains the integrity of the input features, providing richer information support for the subsequent tasks.

2.3 Defect Physical Characteristics Loss

In existing defect detection tasks, model construction mainly relies on data-driven fitting and lacks effective integration of defect mechanism knowledge. Therefore, the model may not fully understand the essential characteristics of defects during training, leading to unreasonable output results and affecting the improvement of detection accuracy.

To address the above issues, a defect physical characteristics loss is designed. This method incorporates key physical features such as the maximum peak-to-valley difference as principle knowledge into the network training process, guiding the model to more effectively learn important features related to MFL defects, thereby improving the accuracy of defect detection. Therefore, the overall loss of the network can be expressed by the following formula:

$$Loss = \frac{1}{N_{cls}}\sum_i L_{cls}(p_i, p_i^*) + \chi\frac{1}{N_{reg}}\sum_i p_i^* L_{reg}(t_i, t_i^*) + \omega L_f \tag{3}$$

where p_i is the predicted category probability, p_i^* is the true category label, N_{cls} is the number of categorization frames, L_{reg} is the CIOU loss, t_i is the predicted box, L_f is the mechanism loss, t_i^* is the true box, N_{reg} is the number of regression box, and χ is the weighting factor.

In particular, the equation of the feature loss L_f is as follows:

$$\begin{cases} \Phi(\mu(\mathbf{z}_i, \mathbf{t}_i)) - \theta \geq 0, \ \alpha_i \geq 0.5 \\ \Phi(\mu(\mathbf{z}_i, \mathbf{t}_i)) - \theta < 0, \ \alpha_i < 0.5 \end{cases} \tag{4}$$

where $\Phi(\cdot)$ denotes the calculation of the maximum peak-to-valley value (MPV), $\mu(\cdot)$ represents the detection network, θ is the MPV threshold, and α_i indicates the confidence score of each predicted region.

The above feature constraints are converted into the following loss function:

$$\begin{cases} L_{low}(\hat{v}, \theta) = \frac{1}{p}\sum_{i=1}^{p}[\sigma(\hat{v}_i - \theta)]^2, & \alpha_i < 0.5 \\ L_{high}(\hat{v}, \theta) = \frac{1}{q}\sum_{i=1}^{q}[\sigma(\theta - \hat{v}_i)]^2, & \alpha_i \geq 0.5 \end{cases} \tag{5}$$

where $\sigma(\cdot)$ corresponds to the ReLU function, $\hat{v}$ denotes the MPV of the predicted region, and p, q are the number of candidate regions with different confidence levels, respectively. The final feature loss is defined as $L_f = L_{low} + L_{high}$. With the above loss mechanism, the model can effectively learn key feature information related to defects, thereby effectively improving the detection performance of the model.

3 Experimental Results and Analysis

3.1 Lab Environment

The hardware environment of the experimental platform: 2 AMD EPYC 7601 2.7 GHz 32-core processors, 1 GeForce RTX 3090 graphics card, 128 GB RAM, LINUX64 Ubuntu18.04 software environment, and the PyTorch framework. The hyperparameter settings for the proposed method are as follows: The optimizer momentum is 0.9. The initial learning rate is 0.005, where its optimization method is MultiStepLR. The size of the network input is resized to 1333 × 888. The batch size is 4.

3.2 Experiment Setting

The measured MFL signals are collected from the experimental field in Fig. 2, which has been located in Tianjin, China. The wall thickness and outside diameter of the experimental pipelines are 12.5 and 325 mm, respectively. The training samples are MFL signals collected from 12-in. pipelines. The number of defects is 200. The test samples include MFL signals from both 12 and 16-in. pipelines. The number of defects is 600.

Fig. 2 Pipeline loop test platform

3.3 Evaluation Metric

We evaluate the performance of the proposed method by the precision and the recall rate. The precision reflects the accuracy of defect detection, and the recall reflects the detection rate of defects. The high recall rate and precision reflect good detection performance. The equations are:

$$P = \frac{TP}{TP + FP}, AP = \int_0^1 P(R)dR \tag{6}$$

$$R = \frac{TP}{TP + FN}, AR = \int_0^1 R(P)dP \tag{7}$$

$$AF - score = (1 + \beta^2) \cdot \frac{AP \cdot AR}{\beta^2 \cdot AP + AR} \tag{8}$$

where TP, TN, FN, and FP denote the true positive, true negative, false negative, and false positive. AR is much more important in defect detection, so β is set to 2.

3.4 Analysis of Results

Several detection methods are used to evaluate the effectiveness of the design method described in this paper. The specific results are shown in Table 1.

The proposed method has achieved optimal performance across all metrics in the table. Specifically, the proposed method demonstrates mAPs of 90.3%, mAP of 92.4%, mAR of 94.7%, and mAF2 of 94.2%. Compared to existing mainstream approaches, the proposed method has shown significant improvements in key evaluation metrics such as accuracy and recall rate. This superior performance is primarily attributed to the method's effective decoupling of classification and regression modules, which enables the extraction of more discriminative features. Furthermore, the loss function designed based on the physical characteristics of defects fully

Table 1 Comparative experiment

Methods	maP_s	mAP	mAR	mAF2
Adaptive faster R-CNN [8]	78.4	82.6	84.5	84.1
TOL-Framework [9]	78.9	83.0	85.4	84.9
YOLO-MF [10]	82.8	84.9	90.5	89.3
YKSB [11]	85.7	87.0	92.1	91.0
PEFNet [12]	86.1	88.6	92.5	91.7
Proposed	**90.3**	**92.4**	**94.7**	**94.2**

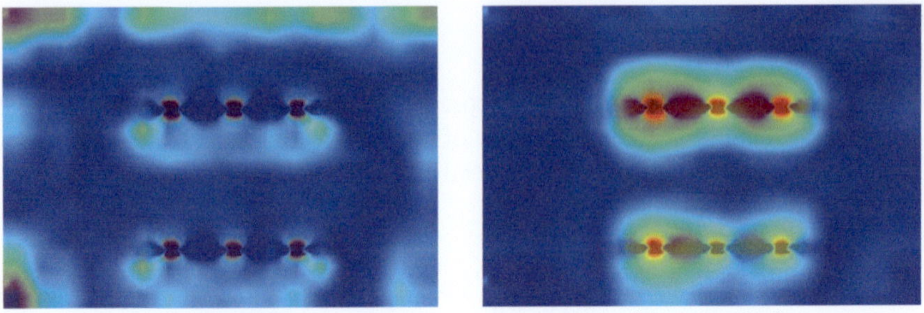

(a) baseline (b) proposed

Fig. 3 Visualization results of attention heat maps

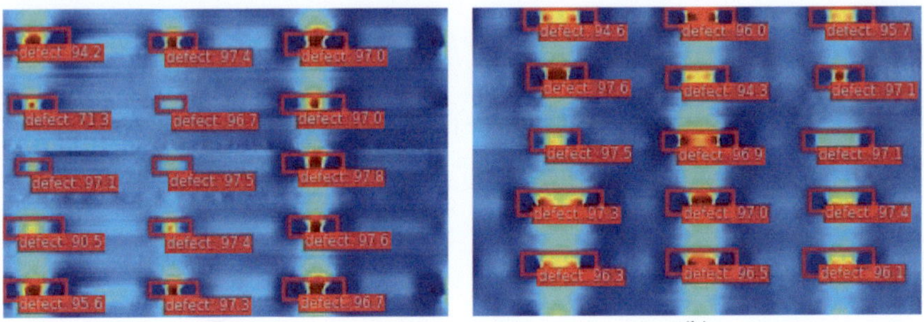

(a) (b)

Fig. 4 Defect detection results

leverages their intrinsic properties, thereby optimizing the model's learning process and enhancing detection accuracy.

In addition, note that a heat map is drawn to illustrate the effectiveness of the design method in this paper, as shown in Fig. 3. As shown in Fig. 3, the design method in this paper can effectively focus on the defect area through the loss and classification regression decoupling module, thereby improving the detection effect of the model.

Finally, the defect detection results are presented, which effectively demonstrate the effectiveness of the method described in this paper. The specific detection results are shown in Fig. 4.

3.5 Ablation Experiment

The ablation experiment results show that introducing CRD and DPC modules can significantly improve the overall performance of the model. When both CRD and DPC modules are presented, the model performance is improved, achieving the

Table 2 Ablation experiment

CRD	DPC	maP$_{s}$	mAP		mAR	mAF2
		86.5	89.0		92.1	91.4
√		88.6	90.9		93.4	92.9
√	√	90.3	92.4		94.7	94.2

best detection results. The experimental results fully demonstrate that the CRD and DPC modules work together to effectively enhance the model's feature extraction capabilities and robustness (Table 2).

4 Conclusion

A MFL Mechanism-Based Intelligent Detection Method for Energy Transmission Systems. The proposed method includes two key innovations: first, a classification-regression decoupling module is designed to optimize the feature extraction process separately for the different requirements of classification and localization tasks, thereby improving the model's discrimination ability and localization accuracy. Second, a defect physical characteristic loss is proposed, which effectively guides the model to focus on key features by incorporating the physical attribute information of defects into the model training, further enhancing the detection effect. Experiments have proved that the proposed method can achieve outstanding detection results in defect detection tasks.

References

1. Zuo F, Liu J, Zhang H, Chen Z, Yan B, Wang L (2025) A complex welding defect detection method based on active learning in pipeline transportation system. IEEE Trans Instrum Meas 74:1–12
2. Liu J, Wen Z, Shen X, Zuo F, Jiang L, Zhang H (2025) Online pipeline weld defect detection for magnetic flux leakage inspection system via lightweight rotated network. IEEE Trans Ind Electron 72(7):7573–7584
3. Rayhana R, Yun H, Liu Z, Kong X (2024) Automated defect-detection system for water pipelines based on CCTV inspection videos of autonomous robotic platforms. IEEE/ASME Trans Mechatron 29(3):2021–2031
4. Jiang L, Zhang H, Liu J, Shen X, Xu H (2025) Mutual supervision of MFL heterogeneous signals for insufficient sample defect detection on pipeline safety operation. IEEE Trans Autom Sci Eng 22:1714–1724
5. Li F, Feng J, Lu S, Liu J, Yao Y (2017) Convolution neural network for classification of magnetic flux leakage response segments. In: 6th data driven control and learning systems (DDCLS), Chongqing, China, pp 152–155

6. Shen X, Liu J, Zhao H, Liu X, Zhang B (2021) Research on multi-target recognition algorithm of pipeline magnetic flux leakage signal based on improved cascade RCNN. In: 3rd international conference on industrial artificial intelligence (IAI), Shenyang, China, pp 1–6
7. Jiang L, Zhang H, Liu J, Shen X, Xu H (2022) THMS-Net: a two-stage heterogeneous signals mutual supervision network for MFL weak defect detection. IEEE Trans Instrum Meas 71:1–9
8. Zhang Y, Zhang Z, Fu K, Luo X (2022) Adaptive defect detection for 3-D printed lattice structures based on improved faster R-CNN. IEEE Trans Instrum Meas 71:1–9
9. Zhong J, Liu Z, Yang C, Wang H, Gao S, Núñez A (2022) Adversarial reconstruction based on tighter oriented localization for catenary insulator defect detection in high-speed railways. IEEE Trans Intell Transp Syst 23(2):1109–1120
10. Ma D, Fang H, Wang N, Zhang C, Dong J, Hu H (2022) Automatic detection and counting system for pavement cracks based on PCGAN and YOLO-MF. IEEE Trans Intell Transp Syst 23(11):22166–22178
11. Shen N, Xv R, Gao Y, Qian C, Chen Q (2024) An improved YOLOv5 model based on feature fusion and attention mechanism for multiscale satellite recognition. IEEE Sens J 24(12):19385–19396
12. Huang L, Huang W, Gong H, Yu C, You Z (2023) PEFNet: position enhancement faster network for object detection in roadside perception system. IEEE Access 11:73007–73023

MIX
Papier aus verantwortungsvollen Quellen
Paper from responsible sources
FSC® C105338

If you have any concerns about our products,
you can contact us on
ProductSafety@springernature.com

In case Publisher is established outside the EU,
the EU authorized representative is:
Springer Nature Customer Service Center GmbH
Europaplatz 3, 69115 Heidelberg, Germany

Printed by Libri Plureos GmbH
in Hamburg, Germany